Business English für die Personalarbeit

René Bosewitz/Annette Bosewitz

Business English für die Personalarbeit

Gesprächsleitfäden, Musterdialoge, Fachvokabular

4. Auflage

Haufe Group
Freiburg · München · Stuttgart

Bibliografische Information der Deutschen Nationalbibliothek

Die Deutsche Nationalbibliothek verzeichnet diese Publikation in der Deutschen Nationalbibliografie; detaillierte bibliografische Daten sind im Internet über http://dnb.dnb.de/ abrufbar.

Print: ISBN 978-3-648-16012-1 Bestell-Nr. 04435-0004
ePub: ISBN 978-3-648-16013-8 Bestell-Nr. 04435-0103
ePDF: ISBN 978-3-648-16014-5 Bestell-Nr. 04435-0153

René Bosewitz/Annette Bosewitz
Business English für die Personalarbeit
4. Auflage, Juli 2022

www.haufe.de
info@haufe.de

Bildnachweis (Cover): ©-strizh-, shutterstock.com

Produktmanagement: Dr. Bernhard Landkammer

Inhaltsverzeichnis

Was Ihnen dieses Buch bietet

Was Sie hier in der Hand halten, ist mehr als ein reiner Ratgeber für die Personalarbeit. Es ist zugleich Sprachtrainer, Wörterbuch, und nützlicher Helfer in Ihrer täglichen Arbeit für alles, was Sie in diesem Bereich auf Englisch formulieren müssen!

Wir führen Sie durch alle für die Personalarbeit wichtigen Themen: Von der Erstellung eines Anforderungsprofils bis zur Einarbeitung eines neuen Mitarbeiters, vom Schreiben eines Arbeitszeugnisses bis hin zum Verfassen einer Kündigung. Und weil Sie ja nicht nur ständig dienstlich unterwegs sind, finden Sie am Ende des Buches viele Tipps, wie Sie in unterschiedlichen Ländern Smalltalk führen.

Nutzen Sie die Online-Arbeitshilfen zum Buch
Auf mybook.haufe.de bieten wir Ihnen praxiserprobte und sofort einsetzbare Mustervorlagen, hilfreiche Gesprächsleitfäden und vieles mehr. Damit sind Sie für alle wichtigen Aufgaben Ihrer täglichen Arbeit gut gerüstet. Zudem finden Sie dort weitere praxisnahe Übungen zu den einzelnen Modulen. Gehen Sie einfach auf mybook.haufe.de und geben Sie dort den Buchcode ein, abgedruckt ist.

Eine kleine Anmerkung in eigener Sache: Die Beispiele aus dem amerikanischen bzw. englischen Rechtsraum sind juristisch betrachtet nicht immer auf das jeweilige, in Ihrem Land geltende Recht anwendbar. Wir können daher hierfür keine Gewähr übernehmen. Bitte prüfen Sie vor wichtigen Entscheidungen immer die rechtliche Lage. Und schließlich: Die in den Beispielen gewählten Namen und Firmenbezeichnungen sind frei erfunden. Mögliche Übereinstimmungen mit real existierenden Personen oder Firmen sind zufällig und nicht beabsichtigt.

Was wir mit diesem Buch-Sprachtrainer-Wörterbuch-Arbeitsmittel beabsichtigen, ist, Sie bei Ihrem Fortkommen zu unterstützen und Ihnen die Arbeit zu erleichtern. Wir hoffen, dass uns dies gelungen ist.

Und Ihnen wünschen wir viel Erfolg!

Heidelberg im Mai 2022

Dr. René Bosewitz und *Annette Bosewitz*

1 Anforderungsprofil und Stellenbeschreibung

Zusammenfassung
Wer einen neuen Mitarbeiter oder eine neue Mitarbeiterin sucht, muss wissen, was diese/r können soll. Dazu ist es äußerst hilfreich, vorab ein Anforderungsprofil zu erstellen und eventuell zusätzlich eine Stellenbeschreibung zu verfassen.

Wir geben Ihnen für diese wichtigen Schritte zu einer optimalen Personalauswahl die sprachlichen Mittel an die Hand und zeigen Ihnen anhand von *Gesprächsleitfäden*, *Beispieldialogen* und mit hilfreichen *Formularen* und *Mustern* aus der Praxis, wie Sie

- den Personalbedarf prüfen,
- ein Anforderungsprofil abstimmen und
- eine Stellenbeschreibung formulieren.

Am Ende dieses Moduls haben wir für Sie nützliche *Fachbegriffe* aufgelistet, zum raschen Nachschlagen und besseren Einprägen. Und anhand einiger *Übungen* können Sie Ihr neu erlerntes Wissen überprüfen (sowie die Lösungen im Anhang dieses Buches nachschlagen).

Die Meilensteine: Anforderungsprofil professionell erstellen

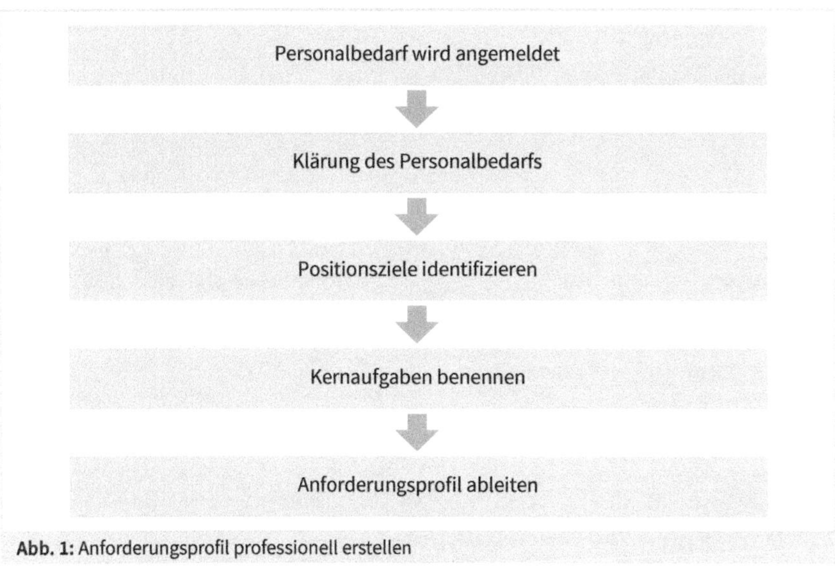

Abb. 1: Anforderungsprofil professionell erstellen

1.1 Personalbedarf und Anforderungsprofil klären

Recruitment is the search for an employee to fill a specific position in the company.	Personalbeschaffung ist die Suche nach einer neuen Arbeitskraft, um eine bestimmte Position im Unternehmen zu besetzen.
This person may be recruited externally by some sort of agency or found within the company. This module deals with some aspects of recruiting, the requirements placed on candidates, an overview of the different recruitment channels and methods as well as job descriptions.	Diese Person kann sowohl extern mittels einer Agentur als auch intern angeworben werden. Dieses Modul beschäftigt sich mit einigen Gesichtspunkten der Personalbeschaffung, den Anforderungen an Bewerber:innen, einer Übersicht zu Wegen und Methoden der Personalbeschaffung sowie Stellenbeschreibungen.

1.1.1 Gesprächsleitfaden: Personalbedarf und Anforderungsprofil klären

Bereiten Sie sich mit dem nachfolgenden Leitfaden auf das Gespräch vor, indem Sie eine Stelle für einen neuen Mitarbeiter bzw. eine neue Mitarbeiterin anfordern.

Wir geben Ihnen Formulierungen an die Hand, wie Sie

- das Thema einleiten,
- den Bedarf erläutern,
- zwischen interner und externer Kandidatensuche abwägen,
- Ihre Präferenzen darlegen,
- die entstehenden Kosten ansprechen,
- die Anforderung an den neuen Mitarbeiter definieren und
- das Vorgehen festlegen.

Introduction of topic	**Einstieg in das Thema**
Actually, Marketing is in need of a …	Die Marketingabteilung benötigt einen …
We're on the lookout for a …	Wir sind auf der Suche nach einem …
At the present time we need to find a …	Im Moment müssen wir einen … finden.
Basically, the company is searching/ looking for a …	Im Grunde sucht die Firma einen …
Explaining department needs	**Bedarf erläutern**
The ideal candidate should be able to …	Der ideale Bewerber sollte in der Lage sein, …
Our department needs someone who …	Unsere Abteilung benötigt jemanden, der …
The following skills are required for this job.	Für diese Stelle sind die folgenden Fähigkeiten erforderlich.

We need someone to fill the position of …	Wir brauchen jemanden, um die Stelle als … zu besetzen.
It has to be a person with leadership potential.	Es muss eine Person mit Führungspotenzial sein.
Considering an internal or external candidate	**Überlegungen zur internen oder externen Stellenbesetzung**
Have you given any consideration as to whether …?	Haben Sie in Erwägung gezogen, ob …
Have you got any preference?	Gibt es eine Präferenz Ihrerseits?
What's your feeling – external or current employee?	Was ist Ihr Gefühl – extern oder eine Mitarbeiterin aus dem Haus?
How do you feel about …?	Was halten Sie von …?
Could it be an external?	Könnte es ein Externer sein?
Stating preferences	**Präferenzen mitteilen**
Well, I would be in favour of …	Also, ich würde es befürworten, wenn …
If I could decide, I would opt for …	Wenn ich entscheiden könnte, würde ich mich für … entscheiden.
Actually, we'd prefer someone who …	Eigentlich bevorzugen wir jemanden, der …
To tell the truth we don't think that …	Um ehrlich zu sein, glauben wir nicht, dass …
Discussing cost and effort	**Kosten und Aufwand ansprechen**
You'll have to think about the costs.	Sie müssen an die Kosten denken.
I'm afraid cost will play a big role.	Ich befürchte, die Kosten werden eine große Rolle spielen.
External recruitment will involve a lot of work.	Externe Personalbeschaffung bringt eine Menge Arbeit mit sich.
We'll have our work cut out.	Wir werden unsere liebe Mühe haben.
Specifying HR requirements for recruiting	**Informationen für HR anfordern**
HR will need the following information …	HR benötigt folgende Daten …
HR has to create the job description	HR muss die Arbeitsplatzbeschreibung erstellen.
Well, HR will have to collect all the necessary information …	Also, HR muss die erforderlichen Informationen zusammenstellen.
We have to optimise the use of all recruitment channels.	Wir müssen die Nutzung aller Beschaffungswege optimieren.

Detailing process	Das Vorgehen festlegen
As soon as we've received the information, we'll define the process.	Sobald wir die Informationen erhalten haben, legen wir den Prozess fest.
We'll prepare the workflow.	Wir bereiten den Ablauf vor.
HR will draw up a timetable	HR erstellt den Terminplan
Finally, we'll invite the potential candidates.	Zum Schluss laden wir die potenziellen Kandidaten und Kandidatinnen ein.

1.1.2 Beispieldialog: Discussing candidate requirements

Background: The dialogue below is between Jim Johnson, the Director of Marketing, and a member of the HR Department, Liz Francis. It is a discussion concerning the recruitment of a person for the Marketing Department.

Jim Johnson: So, basically the company and especially the marketing department is searching for an employee to be trained and eventually take over as section leader and later head of department.

Liz Francis: Okay, and have you given any thought as to whether the candidate should be acquired internally or externally?

Jim Johnson: Yes, we have. And the decision was that we need someone from outside. There's nobody in the company with the relevant skills, as far as we know.

Liz Francis: Recruiting an external will involve a lot of extra cost, advertising, various forms of testing etc.

Jim Johnson: I understand that but we have to get the right person for the job, or we'll have to do it all over again and that will cost more.

Liz Francis: That's true, of course. Well, first of all HR will need the exact requirements for this position so that we can create a job description.

Jim Johnson: That's no problem. We know what we're looking for. A high-flying management trainee with upper-intermediate to advanced competence in two or three languages – German and Spanish at least. And the incumbent should participate in a two-year rotation programme. Background in business administration etc., etc.

Liz Francis: I see where you're going. Well, I'll need all that plus detail in the brief that you send HR. Just to let you know what our process is. As soon as we've received your briefing, we'll prepare the advertisement and decide where to place it. Then we'll nominate a team member to deal with the enquiries. We keep a record of dates when we receive applications from candidates. Then we fix a time slot for interviews. We'll coordinate with you as regards your availability. Next, we'll make a selection and then invite people for a first interview. After the interview we'll have a short list of candidates for second interviews.

Jim Johnson: How many is that normally?

Liz Francis: It depends. Probably two or three. And finally, we'll hopefully have one or two successful applicants that we can contact with a job offer.

Jim Johnson: It all sounds relatively easy.

Liz Francis: Will you be attending the interviews with the short-listed candidates yourself?

Jim Johnson: I'll try to, but it's more important that the present incumbent Ron Stacks is involved. It's his position that will need filling later. It's a question of our successor policy.

1.1.3 Gesprächsleitfaden: Recruitingmethoden klären

Der nachfolgende Leitfaden dient zur Vorbereitung für interne Gespräche über Personalbeschaffungsmaßnahmen. Sie erhalten Formulierungen wie man

- das Ziel einer Besprechung definiert,
- die aktuelle Situation darlegt,
- den Bedarf ermittelt,
- Anforderungen definiert,
- Einwände erhebt und
- Alternativvorschläge unterbreitet.

Defining meeting objectives	Meetingziele definieren
Okay, I've called this meeting to …	Okay, ich habe dieses Meeting einberufen, um …
I've asked you to come here today to …	Ich habe Sie heute hierhergebeten, um …
The goal of today's meeting is to …	Ziel des heutigen Treffens ist es, …
We're meeting today to …	Wir treffen uns heute, um …

Explaining the situation or status	**Den Status quo beschreiben**
The status is this …	Der Stand ist, dass …
The situation at present is …	Die aktuelle Situation ist …
The current position is …	Die momentane Position ist …
Up to now we've done the following …	Bis jetzt haben wir Folgendes unternommen …
Checking needs	**Bedarf definieren**
What do they need in general?	Was benötigen sie generell?
What are the requirements like?	Wie sehen die Anforderungen aus?
What sort of requests do they have?	Was für Wünsche haben sie?
What do they have in mind?	Was schwebt ihnen vor?
Specifying requirements	**Anforderungen klären**
To put it in a nutshell, we require …	Kurz um, wir brauchen …
Well, we certainly expect a …	Also wir erwarten sicherlich …
We need to cover the following requirements …	Folgenden Bedarf müssen wir decken …
One important consideration is …	Eine wichtige Überlegung ist, …
Objecting to a suggestion	**Einem Vorschlag entgegnen**
We can't justify head-hunting.	Das Abwerben von Arbeitskräften können wir nicht rechtfertigen.
We can't explain away such high costs.	Es gibt keine plausible Erläuterung für solch hohe Kosten.
There's no way we can support that idea.	Unter keinen Umständen unterstützen wir diese Idee.
Nobody's going to accept this plan.	Niemand wird diesen Plan unterstützen.
Suggesting another option	**Andere Möglichkeiten vorschlagen**
Actually, I was wondering if we shouldn't …	Eigentlich habe ich mich gefragt, ob wir nicht …
By the way, it crossed my mind that we might …	Übrigens, ich habe mir überlegt, ob …
Well, it occurred to me that we could …	Also, es kam mir der Gedanke, dass wir …
What about the idea of …?	Was ist mit der Idee …?

1.1.4 Beispieldialog: Talking about recruitment methods

Background: The sample dialogue below is a discussion about decisions concerning the placing of advertisements in order to »land the best possible fish« for a job. The discussion is between the head of Human Resources, Sam Bright, and Kim Dean, the responsible recruitment manager.

Sam Bright: Okay, I've called this meeting to discuss the question of the management trainee position we want to fill. You know the incumbent will most probably go to Marketing eventually. Kim, your comments on this one?

Kim Dean: Yes, the status is that I've interviewed the Marketing Director so I know what character is required. Marketing have already sent us a brief so the background material is all available.

Sam Bright: What do they need in general? I mean age limits, formal education, any job experience in this field?

Kim Dean: They've put it all in the documentation but to put it in a nutshell candidates should be between twenty-two and twenty-eight years old and preferably hold a university degree. They are expected to have studied a business-related course, perhaps business administration, business studies or even economics in the field. Marketing also expects several years of job experience preferably between school-leaving and university.

Sam Bright: That might be a bit difficult to cover with everybody running off to their gap years nowadays.

Kim Dean: Well, the objective is to train the incumbent up to the level of department leader.

Sam Bright: Okay, so let's talk about the recruitment channel that would suit this best.

Kim Dean: So, as you know, if we do it externally we have four possibilities – register companies, recruitment agencies, consultancies and the executive search.

Sam Bright: We can't justify »head hunting« for a person entering the company at trainee level, even if he or she's a high potential for a leadership position later. That's too much expense.

Kim Dean: Aren't there any inexpensive ways of recruiting?

Sam Bright: Well, we could contact our usual recruitment agency. They're reasonable.

Kim Dean: Actually, I was wondering if we shouldn't take on the project ourselves. The placing of the adverts, maybe in the Telegraph, Times, and Times Educational. We could cover all the support work and do the timetabling.

Sam Bright: It'll be very time-consuming.

Kim Dean: Well, I've just completed my questionnaire project so I'm on top of things at the moment.

Sam Bright: Okay. Then could you be the project owner for this one? »Management Trainee Marketing«.

1.2 Stellenbeschreibung verfassen

When the requirements for a job are known then responsibilities and tasks should be put into a job description. Existing know-how, experience and skills should be integrated into the job description.	Sind die Anforderungen an einen Arbeitsplatz erarbeitet, fixieren Sie die Verantwortungsbereiche und Aufgaben in einer Stellenbeschreibung. Führen Sie dazu die für die Tätigkeit erforderlichen Fachkenntnisse, Erfahrungen und Fertigkeiten in der Stellenbeschreibung auf.
This description will ensure that the correct information is passed on to the potential future job-holder. The day-to-day work of the job-holder should be very clear to applicants.	Mit dieser Beschreibung wird sichergestellt, dass korrekte Informationen an potenzielle Stelleninhaber:innenweitergegeben werden. Dem:der Bewerber:in muss klar sein, welches seine:ihre täglichen Aufgaben sein werden.

1.2.1 Formular: Job description

I. Background data	
Job title:	Department:
Job number:	Written by:
Today's date:	

II. Job Summary

(Tip: List the more important or regularly performed tasks)

III. Reports to

IV. Supervises

V. Job duties

(Tip: Briefly describe, for each duty, what employee does and, if possible, how employee does it. At the end of each duty show the approximate percentage of time devoted to duty.)

A. Daily duties

	%
	%
	%

B. Periodic duties (Indicate whether weekly, monthly, quarterly, etc.)

	%
	%
	%

C. Duties performed at irregular intervals.

	%
	%
	%

1.2.2 Muster: Job descriptions

Job description 1: Customer Service Clerk	
Job title	Customer Service Clerk Accountant Receptionist Development Engineer *
Department	*
Job summary/general responsibilities	Answers enquiries and gives directions to customers, authorises cashing of customers' cheques, records and returns last charge cards, sorts and reviews new credit applications, works at customer-service desk in department store
Job duties	Authorises cashing of checks: authorises cashing of personnel or payroll cheques (up to a specified amount) by customers wishing to make payment by cheque. Requests identification such as driver's licence from customers, and examines cheque to verify date, amount, signature, and endorsement. Initials check and sends customer to cashier.

* Eine umfangreiche Liste mit Berufs- und Abteilungsbezeichnungen finden Sie im Anhang.

Job description 2: Accounts Office Manager	
Title	Accounts Office Manager
Department	Accounts Payable Department
Definition	This executive has to monitor the procedures of the account office, schedule the activities of staff, liaise with other departments, and work towards the profitability of the company by optimising the use of staff equipment, resources and processes.
Responsibilities	He/she reports to the Administrative Manager for keeping true and fair accounts of all financial transactions between supplier and the company. He/she will pay company invoices on time, support the sales department in their invoicing and present sets of figures to the Admin Manager. He/she will also prepare the company's accounts ledgers on a monthly basis to be inspected by the Financial Director before the annual audit

Job description 3: Director of Studies	
Job title	Director of Studies
Location	Heidelberg, Germany
Reports to	Managing Director, Head Office
Subordinates	Tutors and teachers 1 secretary 2 trainees
Job purpose	to supervise all teaching at customer site and to support teachers
Responsibilities	to check activities of teaching team to prepare materials for support team to perform testing at customers' site to ensure quality

Dos and Don'ts

Die folgenden Dos and Don'ts für Stellenbeschreibungen basieren auf den rechtlichen Anforderungen innerhalb der EU. Sie dürfen somit nicht ungeprüft übernommen werden.

+	Dos
Be specific in the job description e.g. for Director of Studies post it is better to say: test candidates, draw up profiles, liaise with Personnel Development, plan training course.	
Use carefully selected and accurate adjectives, e.g. fast reaction, target-oriented.	
Use clear language which is in general use. Clarity is the key word.	
Keep your language and descriptions neutral to avoid discrimination issues later on.	
List what is only really necessary regarding qualifications for performing the job correctly.	
-	**Don'ts**
Be over-general: »responsibility for quality«	
Use descriptions that are long and wishy-washy: »positive environment that helps learning«.	
Integrate any company-specific jargon or abbreviations. An outsider will be confused and insecure.	
Use any nouns or adjectives which might hint at any sort of discrimination against any group, e.g. able-bodied, young, dynamic, male, female.	
Compile a list of extraordinary qualifications or you'll get candidates who are over-qualified and bored with the job.	

»No go« phrases

»No go« phrase	Reason
We're looking for a dynamic person.	Could be considered as discriminating
We prefer under thirties.	Ageism
We're searching for a specialist in quality.	Is too general, not specific
Are you willing to strengthen collaboration?	Wishy-washy
Have you an analytical approach to ...?	It is too general
Work in our NPI R&D unit.	Abbreviations should not be used. They confuse outsiders
Liaise with our PMA PD staff.	
We expect an able-bodied ...	Discriminating
Youthful staff required.	Ageism

1.3 Vokabeln – Vocabulary input

1.3.1 Aufgabenbeschreibung – Describing job duties

English		German	
Verb (he/she)	**duty**	**Verb**	**Aufgabe**
plans, carries out executes, refines	policies	planen durchführen ausführen weiterentwickeln	Richtlinien
recruits, interviews employs, hires selects, trains dismisses, fires promotes, relocates	employees	anwerben interviewen einstellen einstellen auswählen ausbilden entlassen entlassen befördern versetzen	Arbeitskräfte
conducts introduces explains	new training measures	durchführen einführen erläutern	Schulungsmaßnahmen

English		German	
Verb (he/she)	**duty**	**Verb**	**Aufgabe**
prepares draws up compiles	reports budgets	vorbereiten erstellen zusammenstellen	Berichte Budget
meets with	company representatives/shop stewards	treffen	Unternehmensvertreter:in/Arbeitnehmervertreter:in
represents	the company at hearings and investigations	vertreten	das Unternehmen bei Anhörungen und Untersuchungen
recommends	processes and procedures	empfehlen	Prozesse und Verfahren
negotiates, discusses concludes, signs	contracts with external suppliers/bargaining agreements	verhandeln erörtern abschließen unterschreiben	Verträge mit externen Lieferanten/Tarifverträge
supervises	clerical office staff processes	beaufsichtigen	Büropersonal/Arbeitsprozesse

Nachfolgend finden Sie eine Auflistung wichtiger Fachbegriffe rund um das Thema Personalbeschaffung.

English	German
abbreviation *The abbreviation for Chief Financial Officer is CFO.*	Abkürzung *Die Abkürzung für Chief Financial Officer ist CFO.*
absenteeism	Abwesenheit
accounts ledger	Kontenplan
advertisement	Stellenanzeige
advertising	inserieren, Werbung
age limit	Altersbeschränkung
ageism *Ageism is against the law.*	Altersdiskriminierung *Altersdiskriminierung ist gesetzeswidrig.*
agency	Agentur
applicant	Bewerber:in

English	German
application	Bewerbung
basically	grundsätzlich/im Grunde
brief (n)	Kurzdarstellung
business administration	Betriebswirtschaft
candidate	Kandidat:in
cashier	Kassierer:in
clerical staff	Büropersonal
collective bargaining agreement	Tarifvereinbarung
consideration; give ~ to something *You should give consideration to joining our association.*	etwas in Erwägung ziehen *Sie sollten eine Mitgliedschaft in unserem Verband in Erwägung ziehen.*
cross somebody's mind *It crossed my mind to report the issue to the management.*	jemandem in den Sinn kommen; einfallen *Es kam mir in den Sinn, die Angelegenheit der Geschäftsleitung zu melden.*
degree	Hochschulabschluss
demand *The demand for our products rose last month.*	Nachfrage *Die Nachfrage nach unseren Produkten ist im letzten Monat gestiegen.*
department	Abteilung
duty / task	Aufgabe
economics	Volkswirtschaft
education *I enjoyed a good education.*	Bildung *Ich genoss eine gute Bildung.*
effort	Aufwand
endorsement	Indossament
enquiry *We only had three enquires today.*	Anfrage *Heute hatten wir nur drei Anfragen.*
externally	Extern
favour (v)	bevorzugen
flowchart	Ablaufdiagramm
gap year *After I finished my degree I took a gap year.*	Zeit ohne Beschäftigung/Auszeit *Nach meinem Studium nahm ich mir eine Auszeit für ein Jahr.*

English	German
graduate (n)	Hochschulabsolvent:in
head-hunting	Abwerbung
hearing *The judge decided the case at the hearing.*	Anhörung *Die Richterin entschied den Fall bei der Anhörung.*
hint at something	etwas andeuten
hiring	Einstellen
incumbent	Stelleninhaber:in
internal	intern
interview (n)	Bewerbungsgespräch
job description	Stellenbeschreibung
job offer	Stellenangebot
job title	Stellenbezeichnung
job-seeker	Arbeitssuchender:e
liaise with *You'll need to liaise with three other departments.*	sich abstimmen, Verbindung aufnehmen mit *Sie müssen sich mit drei anderen Abteilungen abstimmen.*
lookout for; to be on the ~ *We're on the lookout for a human resources manager.*	auf der Suche nach … sein *Wir sind auf der Suche nach einer Personalmanagerin.*
management trainee	Trainee
nutshell; put it in a ~ *To put it in a nutshell, we need new products.*	kurz gesagt; auf den Punkt bringen *Um es auf den Punkt zu bringen, wir brauchen neue Produkte.*
opt for *We opted for this applicant because she had the best references.*	sich für etwas entscheiden *Wir haben uns für diese Bewerberin entschieden, da sie über die besten Referenzen verfügte.*
optimise	Optimieren
questionnaire *Fill in the questionnaire and hand it in.*	Fragebogen *Füllen Sie den Fragebogen aus und reichen Sie ihn ein.*
recruit someone *We commissioned a recruitment agency to recruit new employees.*	jemanden anwerben *Wir haben eine Personalagentur beauftragt, neue Arbeitskräfte anzuwerben.*

English	German
recruitment	Personalbeschaffung
recruitment agency	Personalagentur
recruitment channel	Personalbeschaffungsweg
relocation	Umzug
requirement	Anforderung
schedule *Please schedule a meeting with the CEO for next week.*	planen, Termin ansetzen *Bitte planen Sie einen Gesprächstermin mit dem Vorstandsvorsitzenden für nächste Woche.*
search *So far our search for a successor has not been successful.*	Suche *Bislang blieb unsere Suche nach einem Nachfolger erfolglos.*
section leader	Gruppenleiter:in
selection	Auswahl
shop steward *The shop stewards work closely together with the trade union.*	Arbeitnehmervertreter:in *Die Arbeitnehmervertreter arbeiten eng mit der Gewerkschaft zusammen.*
short-list someone	jemanden in die engere Auswahl nehmen
skill *Her negotiating skills are outstanding.*	Fertigkeit, Fachkenntnis, Geschick *Ihr Verhandlungsgeschick ist hervorragend.*
successor	Nachfolger:in
successor policy *It is dangerous to have no successor policy.*	Nachfolgeregelung *Es ist gefährlich, keine Nachfolgerregelung zu haben.*
supervise	Beaufsichtigen
task	Aufgabe
time-consuming	Zeitraubend
timetable	Terminplan
vacant position *You should apply for the vacant position.*	offene Stelle *Sie sollten sich um die offene Stelle bewerben.*
wage rate	Stundenlohn

1.4 Übungen – Exercise time

Task 1: Can you find the opposites of these words?

indecision = ________________

skilful = ________________

irregular = ________________

specialist = ________________

direct = ________________

employment = ________________

Task 2: Definitions! Match vocabulary from the text to the definition.

1. a person who covers or »owns« a job i________________
2. a possibility to do something o________________
3. diplomas, degrees and other paper q________________
4. a period of time after school when people go abroad g________________
5. a formal list of questions on a topic q________________

Task 3: Complete the phrases.

1. to p________________ your product
2. to f________________ a position
3. to p________________ it in a nutshell
4. I'm on ________________ of things now.
5. up to the ________________ of department leader

ARBEITSMITTEL AUF MYBOOK.HAUFE.DE

Auf der Internetseite zum Buch finden Sie weitere Übungen zu diesem Modul. Mit dem Formular Checkliste und Protokoll: Rahmendaten von Bewerbern:innen behalten Sie im Prozedere der Personalauswahl den Überblick. Sie können diese Checkliste speichern, ausdrucken und natürlich auch an den:die zuständigen:e Abteilungsleiter:in per E-Mail versenden.
Mit dem Formular Betriebsratsanhörung holen Sie vom Betriebsrat, falls es einen gibt, die Zustimmung zu der Neueinstellung eines:er Mitarbeiters:in ein. Sie tragen in das Formular die Informationen ein, die nach § 99 Betriebsverfassungsgesetz notwendig sind.
Weitere Stellenbeschreibungen (Leiter:in Vergütungssysteme sowie Leiter:in Weiterbildung und Personalentwicklung) helfen Ihnen bei der Formulierung individueller Stellenbeschreibungen.

2 Erfolgreiche Stellenanzeigen formulieren

Zusammenfassung
In diesem Modul zeigen wir Ihnen, wie Sie eine erfolgreiche Stellenanzeige verfassen. Dazu haben wir für Sie alle wichtigen *Vokabeln*, *Fachausdrücke* sowie zahlreiche *Formulierungen* und *Textbausteine* zusammengestellt. Weitere Muster für Stellenanzeigen können Sie sich unter mybook.haufe.de ansehen und herunterladen.

Die Meilensteine: Professionelle Stellenanzeigen entwickeln

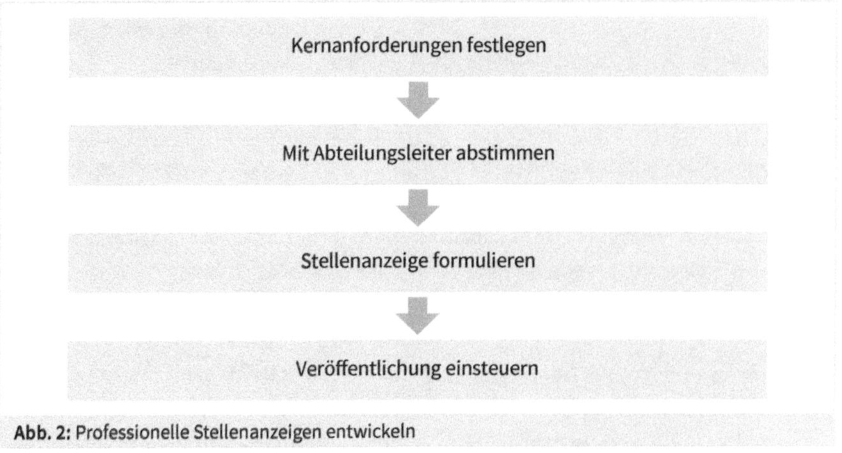

Abb. 2: Professionelle Stellenanzeigen entwickeln

2.1 Schwerpunkte der Stellenanzeige

An employer may not use biased language in advertising for jobs. If an advert discriminates against or discourages certain groups from applying for a job, it will be considered illegal. This includes discrimination based on sex and marital status, race, disability, sexual orientation, religion, belief, and age.	Die Sprache, die Arbeitgeber in ihren Stellenanzeigen verwenden, darf weder voreingenommen noch bewertend sein. Sobald eine Anzeige bestimmte Gruppen diskriminiert oder sie abschreckt, sich zu bewerben, gilt diese als gesetzeswidrig. Hierzu zählt eine Diskriminierung aufgrund der Merkmale Geschlecht, Familienstand, Rasse, Behinderung, sexuelle Neigung, Religion, Glaube oder Alter.
If the advert contains a phrase like »salesman needed« this would imply that women will not be considered for the position and is therefore a breach of the law concerning gender equality. The term »salesperson« should have been used in the advert.	Enthält eine Anzeige Formulierungen wie beispielsweise »Verkäufer gesucht« wird dies derart ausgelegt, dass Frauen für diese Stelle nicht berücksichtigt werden. Sie verstößt folglich gegen das Allgemeine Gleichbehandlungsgesetz (AGG). Der Begriff »Verkäufer:in« hätte in der Anzeige verwendet werden sollen.
A further example might be »single professional teacher required«. This would discriminate on the basis of marital status as it would appear to exclude married people from the chance of getting the post.	Ein weiteres Beispiel hierfür ist »alleinstehender:e, professioneller:e Lehrer:in gesucht«. Hier liegt eine Diskriminierung aufgrund des Familienstands vor, da es scheint, dass verheiratete Menschen von der Möglichkeit, diese Stelle zu bekommen, ausgeschlossen werden.

2.1.1 Gesprächsleitfaden: Inhalt der Stellenanzeige diskutieren und festlegen

In dem nachfolgenden Leitfaden werden folgende Gesprächssituationen behandelt:

- Schwerpunkte der Stellenanzeige klären
- seine eigene Meinung äußern
- Vermutungen äußern
- Vorschläge unterbreiten
- Empfehlungen geben

Asking what should be included in general	Nach den Schwerpunkten der Stellenanzeige fragen
What should we include if we ...?	Was sollten wir mit aufnehmen, wenn wir ...?
What do you think we should put in there?	Was denken Sie, sollten wir angeben?
What ought to be the priority of the advert?	Was sollte der Schwerpunkt der Stellenanzeige sein?
What is a »must« for this ad?	Was muss auf jeden Fall in dieser Stellenanzei-ge enthalten sein?
Giving your opinion	**Die eigene Meinung äußern**
As far as I'm concerned ...	Was mich betrifft, so ...
Well, I reckon that ...	Also, ich glaube, dass ...
If you ask my opinion ...	Wenn Sie nach meiner Meinung fragen, ...
In my opinion ...	Meiner Meinung nach, ...
If you ask me ...	Wenn Sie mich fragen, ...
It seems to me that ...	Mir scheint, dass ...
I'd say that ...	Ich würde sagen, dass ...
Making assumptions	**Vermutungen äußern**
I suppose the candidate will ...	Ich denke, der Kandidat wird...
I expect the candidate to ...	Ich erwarte, dass die Kandidatin ...
He'll probably go for ...	Er entscheidet sich wahrscheinlich für ...
She'll probably go for the strategy of ...	Sie wird wohl die Strategie des ... wählen.
Later he'll decide for ...	Später wird er sich für ... entscheiden.
She may opt for ...	Sie wird sich vielleicht für ... entscheiden.
Making suggestions	**Vorschläge unterbreiten**
Okay, so we should create ...	Okay, wir sollten ... erstellen.
Perhaps we ought to put together ...	Vielleicht sollten wir .... zusammenstellen.
Let's think up a ...	Denken wir uns einen ... aus.
Might it be an idea, if ...?	Wäre es eine Idee, wenn ...?
I was wondering if we could ...	Ich habe mich gefragt, ob wir ...
Have you ever thought of ...?	Haben Sie jemals daran gedacht ...?
Why don't we ...?	Warum machen wir nicht ...?
Wouldn't it be better to ...?	Wäre es nicht besser, wenn ...?

Making recommendations	Empfehlungen geben
I'd recommend him to ...	Ich würde ihm empfehlen, dass er ...
If you ask me she should ...	Wenn Sie mich fragen, dann sollte sie ...
If I were him I'd ...	Ich an seiner Stelle würde ...
The candidate might like to ...	Die Bewerberin würde vielleicht gerne ...
Wouldn't it be better for her to ...?	Wäre es nicht besser für sie, wenn ...?
Wouldn't it be wiser to ...?	Wäre es nicht klüger, wenn ...?
Hadn't he better try to ...?	Sollte er nicht besser versuchen, ...?
The best thing to do is ...	Das Beste ist, ...

2.1.2 Beispieldialog: About the content of an advert

Background: Discussion between the recruiter in the personnel department, Frank Dean, and the head of the Marketing Department, Jim Johnson.

Frank Dean: So in your opinion what should we include if we want to get a really good candidate for this position?

Jim Johnson: As far as I'm concerned there are several items of information that should be available to the applicant. Perhaps most important is the reputation of the company. Something should be said about that.

Frank Dean: Okay. So we should create a one or two-line slogan or statement on that. It ought to be easy enough.

Jim Johnson: Then we would state clearly which department the applicant is intended for. You know R&D, Purchasing, Sales, etc. And what type of climate exists there, what sort of psychological environment.

Frank Dean: Oh, that's interesting. You mean »is the climate friendly, hierarchical or multi-disciplinary?«

Jim Johnson: That's right. And the candidate might like to check on the success of its management team over the years.

Frank Dean: And what about the tasks themselves?

Jim Johnson: Yes, the exact nature of all duties expected. But, of course, reduced down to a single phrase. After all it's restricted to an advert.

Frank Dean: Well, I suppose the candidate will be in contact with other departments. Shall we try to include his or her degree of involvement with other sections and departments?

Jim Johnson: Yes, of course, and especially the links to the superior and his or her superior's boss. And what or who the candidate would be responsible for. And by the way we have to agree on the remuneration package and what we want to advertise publicly, at least in the UK.

Frank Dean: You mean bonuses, company car, entertainment allowance, health insurance and so on?

2.2 Eine Stellenanzeige effizient erstellen

2.2.1 Checkliste: Looking at job adverts

When writing an advert make sure that it complies with the legal requirements of equal opportunities.	Achten Sie beim Erstellen einer Anzeige darauf, dass die Anzeige die rechtlichen Vorgaben zur Gleichbehandlung berücksichtigt.
Phrases such as »young specialists« or »motivated young employee« indicate ageism.	Formulierungen wie »Junger:e Experte:in« oder »Motivierter:e junger:e Mitarbeiter:in« sind Indikatoren für Altersdiskriminierung.
Put your adverts in places where men and women, young and old will see them.	Schalten Sie Ihre Anzeigen dort, wo Männer, Frauen, junge sowie alte Menschen sie sehen können.
Don't present men and women in stereotyped situations.	Zeigen Sie Männer und Frauen nicht in klischeehaften Situationen.
Be careful with word-of-mouth advertising – it excludes people.	Seien Sie vorsichtig mit Mundpropaganda - sie schließt Personen aus.

2.2.2 Textbausteine: Phrases for adverts

Der folgende Leitfaden bietet Ihnen wichtige Formulierungen für die Erstellung von Stellenanzeigen.

Statement of advert status, goal and location	**Aussagen in der Anzeige über Status, Ziel und Standort**
We currently have two vacancies.	Zurzeit haben wir zwei offene Stellen.
At the moment we can offer two jobs.	Im Moment können wir zwei Stellen anbieten.
There are two positions available.	Es gibt zwei verfügbare Stellen.
There are two jobs on offer.	Zwei Stellen werden angeboten.
We are located in …	Unser Firmensitz ist in …
The headquarters is situated in …	Unser Hauptsitz ist in …
Job title	**Berufsbezeichnungen**
Auf mybook.haufe.de finden Sie weitere Berufsbezeichnungen.	
Client relationship manager	Kundenbetreuer:in
Clerk	Sachbearbeiter:in
Book-keeper	Buchhalter:in
Sales representative	Außendiensmitarbeiter:in; Handelsvertreter:in
Auditor	Wirtschaftsprüfer:in
Safety officer	Sicherheitsbeauftragter:e
Draftsperson	Technischer:e Zeichner:in
Engineer	Ingenieur:in/Techniker:in
Mechanical engineer	Maschinenbauer:in
Description of tasks	**Aufgabenbeschreibung**
You will develop and maintain client relationships …	Sie entwickeln und pflegen Kundenbeziehungen.
You will be developing …	Sie entwickeln …
You will be raising client awareness.	Sie steigern das Kundenbewusstsein.
You will be working closely with …	Sie arbeiten eng zusammen mit …
Your tasks include …	Zu Ihren Aufgaben zählen …
You will be responsible for …	Sie sind verantwortlich für …
You will lead a small team.	Sie leiten ein kleines Team.
Working in the accounts department, you will …	Bei Ihrer Arbeit in der Buchhaltung werden Sie …
This role will include …	Diese Funktion umfasst …

Skills and soft skills/attitude	Fertigkeiten und Soft Skills
We are looking for self-starters.	Wir suchen Mitarbeitende mit Eigeninitiative
We need independent thinkers.	Wir benötigen eigenständige Denker:innen.
Entrepreneurial thinking and attitude are required.	Unternehmerisches Denken und Einstellung werden vorausgesetzt.
You must be open to new ideas.	Sie müssen für neue Ideen offen sein.
You should be willing to learn.	Sie sollten lernwillig sein.
You need the ability to learn/acquire new skills.	Sie müssen in der Lage sein, neue Fähigkeiten zu lernen/sich anzueignen.
Communication skills are a plus in this job.	Kommunikationsfähigkeiten sind ein Plus bei dieser Arbeit.
Presentation skills are a must.	Präsentationskenntnisse sind ein Muss.
Fluency in French/German/Spanish …	Fließend in Französisch/Deutsch/Spanisch.
Good knowledge of English.	Gute Englischkenntnisse.
Working knowledge of Russian.	Grundkenntnisse in Russisch
Formal qualifications	**Formale Qualifikationen**
You should possess a degree or equivalent.	Sie sollten über einen Hochschulabschluss oder eine vergleichbare Ausbildung verfügen.
Post-graduate degree in Business Administration required.	Diplom in Betriebswirtschaft ist erforderlich.
Diploma or degree is an advantage.	Diplom oder Hochschulabschluss von Vorteil.
We expect qualifications in IT, Word, Excel and PowerPoint.	Wir erwarten Qualifikationen in IT, Word, Excel und PowerPoint.
MBA is the minimum.	MBA ist Mindestvoraussetzung.
German-speaking graduate.	Deutschsprachiger Hochschulabsolvent.
Educated to degree level.	Ausbildung mit Hochschulabschluss.
Work experience	Berufserfahrung
Work experience an advantage	**Berufserfahrung von Vorteil**
Experience in purchasing/finance	Erfahrung im Einkauf/Finanzwesen.
Experience abroad an advantage	Auslandserfahrung von Vorteil.
Experience in a cross-cultural/multinational company necessary	Erfahrung in einem interkulturellen/multi-nationalen Unternehmen erforderlich.

Statement on salary	Aussage über das Gehalt
Salary ca. £ 22 – 26 kpa	Jahresgehalt etwa 22.000 £ – 26.000 £
Remuneration ca. £ 22,000 p. a.	Vergütung ca. 22.000 £ pro Jahr.
Starting salary ca. £ 22,000 p. a.	Anfangsgehalt von ca. 22.000 £ pro Jahr.
You will receive a competitive salary.	Sie erhalten ein konkurrenzfähiges Gehalt.
We offer an attractive salary package.	Wir bieten ein attraktives Paket.
Benefits/perks include pension scheme, company car …	Sonderleistungen/freiwillige Sozialleistungen umfassen Altersversorgung, Firmenwagen …
You receive 26 days paid leave/holiday.	Sie erhalten 26 Tage bezahlten Urlaub.
Brief description of company	**Kurze Beschreibung des Unternehmens**
We're one of the UK's leading companies in …	Wir zählen zu den führenden Unternehmen in England in …
We're leading edge in …	Wir sind führend in …
XYZ is a pioneer in the field of …	XYZ ist ein Vorreiter im Bereich …
We are the market leader in …	Wir sind Marktführer in …
We stand at the forefront in the area of …	Wir gehören zu den Besten im Bereich …
Our core business is …	Unser Kerngeschäft ist …
We are a well-known service provider of …	Wir sind ein bekannter Dienstleister für …
Our company was founded xx years ago.	Unser Unternehmen wurde vor xx Jahren gegründet.
We have been operating since …	Wir sind seit … tätig.
Contact	**Ansprechpartner:in**
Please contact …	Bitte wenden Sie sich an …
Please write to/email …	Bitte schreiben Sie an/senden Sie Ihre E-Mail an …
Send your CV with covering letter to …	Senden Sie Ihren Lebenslauf zusammen mit Ihrem Anschreiben an …
Send your application quoting the reference no. to …	Senden Sie Ihre Bewerbung unter Angabe der Referenznr. an …
All applications should be received by …	Alle Bewerbungen müssen bis spätestens zum … eintreffen.
Application deadline is …	Stichtag für Ihre Bewerbung ist …

2.2.3 Muster: Stellenausschreibungen – Job advert

Beispiel 1: Graduate Trainee

Job advert	Annotation
What are you	
One Graduate contains	challenging question
Talent 100 % – Drive 100 % – Ambition 100 % – Flair 100 % – Stamina 100 %	
of the total requirement	
made of?	motivating and positive Header
Great prospects for German-speaking graduates Wimbledon – attractive salaries	interesting
Here in the UK, in this challenging economic climate, we are one of UK retailing's great success stories. Uncompromising on quality, we expect the same of our graduate trainees.	emotional flavour
You should be hungry for success, eager to develop your skills and leadership qualities with an international market leader.	motivation
Educated to degree level and ideallyin a business related subject, you will be based at our UK head office in Wimbledon. Working in our purchasing department dealing with many of our Pan-European suppliers, you will need fluent German as well as English and be an effective communicator.	skills
In return for your drive, commitment and ability to deliver at pace, we can offer you serious opportunities for real progress in food buying, supply chain and advertising.	attitude
Please e-mail your CV and covering letter quoting reference TT/19 to: ...	

Beispiel 2: Management Trainee

Management Trainee Needed

We are a market leader in the UK in the field of sports equipment and clothing. It is our vision to support health and sports – a healthy body, a healthy mind and soul. We have a turnover of £80 million p. a. and a workforce of 1800.

Your tasks

- learning and mastering the main business processes
- job rotation in two line departments and four staff departments
- involvement in two major projects
- leading a small project team
- assessing sales potential

Your qualifications
• potential as a strategic thinker • MBA • experience abroad • good knowledge of German and Spanish
Your benefits!
• higher management position after training • starting salary ca. £ 28,000 p. a., pension scheme • company car, relocation expenses, other perks

Beispiel 3: Client relationship manager

We currently have two vacancies in our busy business development team, which is based in the City of London, working in the financial services industry.

Client relationship manager (permanent role)

Salary c. £ 22 – 26 kpa (depending on experience)

Within this role, you will be developing and maintaining relationships with existing and prospective clients and members in the major financial institutions. You will also be raising awareness of the Institute, developing new business opportunities, and working closely with our network of regional committees.

Administrator/Trainee client relationship manager

(12 month contract to cover maternity leave)

Salary c. £ 22 – 26 kpa (depending on experience)

This role combines providing administration support for the business development team and senior managers and is an opportunity to train as a client relationship manager. You will be producing reports, mailings and letters, dealing with calls from corporate clients, arranging travel and itineraries, diary management and covering the office as the team are often out visiting clients.

For both roles we are looking for people who are self-starters with:

The right personal stature and skills to represent the Institute

The ability to learn

A degree or equivalent

IT-Word, Excel and PowerPoint skills

Previous work experience would be an advantage

The Securities & Investments Institute, a registered charity, is the largest and most widely respected professional body for those who work in the UK securities and investment industry. Its aim is to promote higher standards of competence and integrity among its 40,000 members and industry, and as the main examining body for the industry it offers its internationally recognised examinations globally.

There is a competitive remuneration package, which includes a non-contributory pension, private medical insurance, 26 days paid leave per annum, insurances and a paid travel ticket after one year's employment.

Please e-mail your CV and covering letter stating the reference no. of the job you are applying for and your salary expectations to …, head of HR, at recruitment@xxx.co.uk

The closing date for applications is 9 am Tuesday 1st September 2010.

Dos and Don'ts

+	Dos
Do place adverts where they will reach men and women and different age groups.	
-	**Don'ts**
Don't use terms like: »young specialists« or »experienced and mature person«. These infringe the law concerning ageism.	
Don't show men and women in stereotyped roles.	
Don't select employees by word of mouth. It may exclude candidates of a certain sex or race.	
Don't dismiss overseas qualifications. State that, if a certain qualification is needed for a job, then an overseas qualification may be as acceptable as a national one.	

Say	Don't say
We are looking for specialists in …	We are looking for young specialists.
The applicant should have experience of … / The applicant should have several years of job experience.	The applicant should be experienced and mature.
Usual years of experience required.	You must not be older than …
Male or female personal assistant required.	Female assistant required.

2.3 Vokabeln – Vocabulary input

Nachfolgend finden Sie eine Auflistung aller wichtigen Fachbegriffe aus dem Modul in alphabetischer Reihenfolge.

English	German
age	Alter
ageism	Altersdiskriminierung
assess *On what basis will you assess her?*	bewerten *Auf welcher Grundlage bewerten Sie sie?*
assumption	Annahme
available	verfügbar
awareness	Bewusstsein
believe *I believe in your skills.*	glauben, vertrauen *Ich vertraue Ihren Fertigkeiten.*
benefits	Arbeitgeberleistungen
biased *The division leader is biased about his employees.*	befangen; parteiisch; voreingenommen *Die Bereichsleiterin istihren Mitarbeitern gegen-über voreingenommen.*
communication skills	kommunikative Fähigkeiten
company car	Firmenwagen
competitive	Konkurrenzfähig
concerning	bezüglich, betreffend
core business	Kerngeschäft
corporate clients	Firmenkunden
covering letter *Never send a CV without a covering letter.*	Anschreiben *Schicken Sie einen Lebenslauf nie ohne An-schreiben.*
deadline *Our deadline for the report is already next week.*	Abgabefrist *Unsere Abgabefrist für den Lebenslauf ist be-reits nächste Woche.*
degree	Universitätsabschluss
disability	Behinderung
discourage (v)	entmutigen
discriminate	Diskriminieren

English	German
dismiss *Don't dismiss her proposal.*	verwerfen; ablehnen *Verwerfen Sie ihren Vorschlag nicht.*
entertainment allowance	Bewirtungsspesen
environment	Umwelt
exclude	ausschließen
gender	Geschlecht
graduate *She is a graduate of a technical university.*	Absolvent *Sie ist Absolventin einer Technischen Universität.*
headquarters	Hauptsitz
hierarchical *Our company has a strictly hierarchical structure.*	hierarchisch *Unsere Firma hat eine strikt hierarchische Struktur.*
holiday	Urlaub
illegal	gesetzeswidrig
imply *That implies that we invest.*	bedeuten; voraussetzen *Das setzt voraus, dass wir investieren.*
include	einschließen
job advert/advertisement	Stellenanzeige
leading edge *We are leading edge in printing press technology.*	führend in etwas sein *Wir sind führend in der Druckmaschinentechnologie.*
leave	Urlaub
link	Verbindung
marital status	Familienstand
market leader	Marktführer
master the business process	den Geschäftsprozess beherrschen
mature *We need a mature employee for this position.*	reif *Für diese Stelle brauchen wir einen reifen Mitarbeiter.*
multi-disciplinary	multi-disziplinär
offend	verletzen
overseas qualifications	ausländische Qualifikationen
pension scheme	Altersvorsorge

English	German
perks *Perks include company car and private medical insurance.*	Sonderzulagen *Zu den Sonderzulagen zählen Firmenwagen und private Krankenversicherung.*
position	Stelle
prospects	Aussichten
raise awareness	Bewusstsein schärfen
recommendation	Empfehlung
relocation expenses	Umzugskosten
remuneration *Your remuneration will be EUR 90,000 p. a. (per annum).*	Vergütung *Ihre jährliche Vergütung wird 90.000 EUR betragen.*
reputation	Ruf
respective	jeweilig
restrict *We should restrict access to crictical information to key employees.*	beschränken *Wir sollten den Zugriff auf sensible Informationan auf wichtige Mitarbeiter bzw. Mitarbeiterinnen beschränken.*
self-starter	Mitarbeiter:in mit Eigeninitiative
skills	Fähigkeiten
slogan	Slogan
statement	Aussage
strategic thinker	Strategischer:e Denker:in
superior (n)	Vorgesetzter:e
supply chain	Lieferkette
turnover *Our turnover has risen year-on-year.*	Umsatz *Unser Umsatz ist im Vergleich zum Vorjahr gestiegen.*
uncompromising	kompromisslos
vacancies	offene Stellen
word of mouth advertising	Mund-zu-Mund-Propaganda
workforce *We have a workforce of 600 employees.*	Belegschaft *Unsere Belegschaft zählt 600 Mitarbeiter und Mitarbeiterinnen.*

2.4 Übungen – Exercise time

Task 1: Which of the words in the brackets is correct in these sentences?

1. Women would not be (considered) (requested) for this position.
2. The candidate might like to (prove) (see) the company for himself.
3. (Mouth-to-mouth) (Word-of-mouth) advertising can be quite risky.
4. He is educated to (grade) (degree) level.
5. His task will include assessing sales (possibilities) (potential)

Task 2: What is the opposite of the following words?

1. ability = ________________
2. experienced = ________________
3. advantage = ________________
4. available = ________________
5. include = ________________
6. dependent = ________________
7. purchasing = ________________

Task 3: Find the English for the German words.

1. We are a well-known ________________ (*Dienstleister*).
2. If you want my ________________ (*Meinung*) we should invest.
3. Our ________________ (*Hauptsitz*) is located in Cologne.
4. ________________ (*Gesammelte Berufserfahrung*) will be an advantage.
5. Good ________________ (*Englischkenntnisse*) is expected.
6. ________________ (*Stichtag*) for you application is Monday, 1 December.

ARBEITSMITTEL AUF MYBOOK.HAUFE.DE

Auf der Internetseite zum Buch finden Sie weitere Übungen zu diesem Modul sowie zwei weitere Muster für Stellenanzeigen für Vertriebsleiter:innen und Maschinenbauer:innen.

3 Bewerbungen analysieren

Zusammenfassung
Wir zeigen Ihnen in diesem Modul, wie Bewerbungen aufgebaut sind, welche Arten von Lebensläufen je nach Herkunftsland üblich sind und was diese Formen jeweils mitteilen. Bei der Analyse und Bewertung der Lebensläufe und Anschreiben unterstützen wir Sie mit *Checklisten* und *Mustern*, damit Sie bei der Durchsicht der Bewerbungen alles im Blick haben.

Die Meilensteine: Bewerbungen bewerten und eine Auswahl treffen

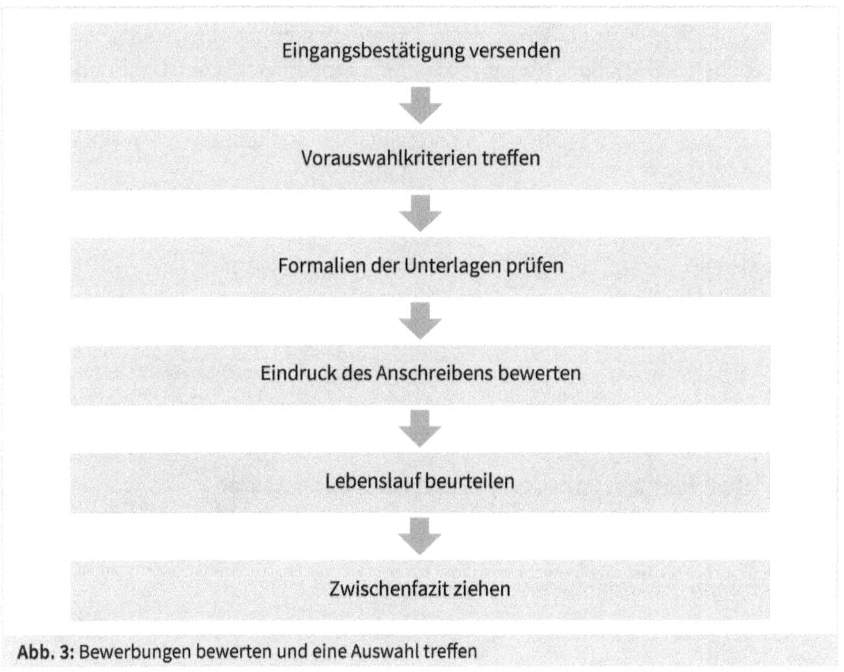

Abb. 3: Bewerbungen bewerten und eine Auswahl treffen

3.1 Lebensläufe verstehen und bewerten

The curriculum vitae (CV) is the first link between the candidate and his prospective employer. The applicant wishes to sell himself to his best advantage, the company wants the best employee for the job. There are several rules that should be followed.

The CV means »the course of your professional life«. In the USA it is called a resume. It offers prospective employers enough information to want to invite the best candidates to an interview.

In general there are three types of CV
- chronological
- functional
- one-page summary

Der Lebenslauf ist der erste Kontakt zwischen Bewerber:in und dem zukünftigen Arbeitgeber. Bewerber:innen möchten sich möglichst vorteilhaft verkaufen, während Unternehmen nach der besten Besetzung für die Stelle suchen. . Es gibt einige Regeln, die dabei zu beachten sind.

CV bedeutet »der Verlauf des Berufslebens«. In den USA wird dies auch »Resume« genannt. Es bietet einschlägigen Arbeitgebern genügend Informationen, um die besten Bewerber:innen zu einem Vorstellungsgespräch einzuladen.

Grundsätzlich gibt es drei Arten von Lebensläufen
- chronologische
- funktionsbezogene
- 1-seitige Zusammenfassungen

3.1.1 Übersicht: Verschiedene Typen von Lebensläufen

Die Tabelle zeigt Ihnen, welche Arten von Lebensläufen üblich sind und warum Bewerber:innen ihren Lebenslauf in dieser Form aufbauen.

Type	Reason	Form
Chronological	Candidate wants to apply for job in the same field.	It gives candidate's job history in *reverse* chronological order.
Functional	First-time job-hunting. Change of career direction: Candidate wants to emphasise transfer skills.	It underlines candidate's main skills.
One-page	Candidate wants to apply for job in the same field.	Summary of candidate's career and track record.

3.1.2 Übersicht: Kulturelle Unterschiede in Lebensläufen

In den verschiedenen Kulturen und Ländern haben sich unterschiedliche Vorstellungen entwickelt, was ein Lebenslauf beinhalten sollte. Für Sie ist es wichtig, diese Unterschiede zu kennen, um zu einer angemessenen Einschätzung der Lebensläufe zu gelangen.

Country	Cultural difference
Different countries have their own ideas about what should be included in CVs. Here is a short overview of what to watch for:	
USA	There are strict rules on the type of information an employer is permitted to know about the candidate, one-page CV is standard.
Europe	Often includes a photo, schooling and higher education integrated, 4 or 5 page CV with detailed information.
Japan, Far East	Only graduates of special universities can apply to certain companies.
France	Handwritten covering letters are expected and can be sent to graphologists for analysis.
United Kingdom	In addition to the CV the covering letter is important. It should contain the contact details of 3 referees who will then be contacted privately by the prospective employer. Open testimonials are not seen as very useful.

3.1.3 Checkliste 1: Einen chronologischen Lebenslauf prüfen

Stating personal details	Persönliche Angaben machen	[x]
Name/surname/family name	Nachname	
First name/Christian name/given name	Vorname	
Address/contactable at …	Adresse/zu erreichen unter …	
email	E-Mail	
Date of birth/DOB/born	Geburtsdatum/geboren am	
Work/job experience	**Arbeits-/Berufserfahrung**	
Engaged on a part-time basis as …	Teilzeitbeschäftigt als …	
Participated successfully in	Erfolgreich teilgenommen an …	
Developed a new system for …	Ein neues System für … entwickelt	
Successfully restructured the department of …	Die Abteilung für … erfolgreich umstrukturiert	

Was solely responsible for ...	Ich war allein verantwortlich für ...	
Intensively coached the staff in ...	Die Belegschaft eingehend in ... betreut	
Was actively involved in ...	Aktiv beteiligt in ...	
Formal qualifications	**Formale Qualifikationen**	
Passed my exams in ...	Meine Prüfungen in ... erfolgreich abgeschlossen.	
Was awarded a university degree in ...	Einen Hochschulabschluss in ... erhalten haben.	
Obtained my GCE/A levels/O levels/baccalaureate/diploma/certificate in	Mein Fachabitur/Abitur/Mittlere Reife/Hochschulreife/Diplom/Abschluss in ... erhalten.	
Completed my traineeship/apprenticeship in ...	Meine Berufsausbildung/Lehre als ... abgeschlossen	
Passed the exam of the Chamber of Commerce, University of	Das Examen der Handelskammer, Universität in ... bestanden.	
Passed the ... exam of ... at advanced/intermediate level.	Erfolgreich die ... Prüfung als ... für fortgeschritten/mittlere Kenntnisse abgelegt.	
Was awarded my degree in ... from the University of	Den Hochschulabschluss in ... von der Universität ... zuerkannt bekommen.	

Bei der Aufzählung der Berufserfahrung ist es nicht notwendig, das Wort »Ich« ständig zu wiederholen. Im Lebenslauf sollten vielmehr dynamische Verben eingesetzt werden, die Handeln und Erfolg demonstrieren. Als Zeitform im englischen CV wird überlicherweise das Simple Past verwendet..

3.1.4 Checkliste 2: Funktionsbezogenen Lebenslauf prüfen

Item	Thema	[x]
Employment	**Beschäftigung**	
Was employed as ...	War beschäftigt als ...	
Worked as ...	Arbeitete als ...	
I worked for XY for xx years	Ich arbeitete für XY für XX Jahre	
My previous employer was ...	Mein früherer Arbeitgeber war ...	
Worked from xx to xy for XX	Arbeitete in der Zeit von xx bis xy für XX	
Was an employee of XX from xx to xy	War in der Zeit vom xx bis xy Angestellter bei XX	

Item	Thema	[x]
Career profile	**Berufliches Profil**	
My interest in … led me to …	Mein Interesse an XX führte mich zu …	
My special training in … resulted in …	Meine spezialisierte Ausbildung in XX resultierte in …	
My career has advanced/developed because …	Meine Karriere ging steil nach oben/hat sich vorteilhaft entwickelt, weil …	
My tasks/duties included …	Meine Funktionen als/Aufgaben beinhalteten …	
Was solely responsible for …	Ich war allein verantwortlich für …	
Was solely in charge of …	Ich war allein zuständig für …	
Was successfully involved in several projects including one which …	War erfolgreich in mehreren Projekten tätig, eingeschlossen ein Projekt, welches …	
Managed to get promotion to …	Gelang mir die Beförderung zum …	
Was promoted to …	Wurde ich zum … befördert.	
Personal profile	**Persönliches Profil**	
I've always been considered a …	Ich wurde immer für … gehalten.	
I've always been regarded as …	Ich bin immer als … angesehen worden	
I like to lead people.	Ich führe gerne Menschen …	
I see my strengths in the managing of projects.	Mein Stärken sehe ich im Leiten von Projekten.	
I enjoy motivating my team.	Ich motiviere gerne mein Team.	
I am computer literate.	Ich kenne mich mit Computern gut aus.	
I am up to date with …	Ich bin auf dem Laufenden bei …	
I have hands-on/expert knowledge of …	Ich habe in der Praxis erprobtes Wissen/Fachkenntnisse in …	
Other information	**Sonstige Informationen**	
Have a clean driving licence.	Im Besitz eines Führerscheins ohne Verkehrsregistereinträge sein	
Have no convictions.	Keine Vorstrafen haben	
Have 10 years' driving experience.	10 Jahre Erfahrung als Fahrer haben	
Have no problem in relocation/changing/moving to …	Umzugsbereit/räumlich flexibel sein	
Am free to move/relocate/transfer to …	Jederzeit umziehen/wechseln können	
Speak fluent English/Russian/Chinese …	Fließend in Englisch/Russisch/Chinesisch	
Have an intermediate knowledge of …	Mittelstufenkenntnisse in … haben	

3.1.5 Muster 1: Chronological CV (reverse)

Curriculum Vitae		
Personal Details		
Name:	Peter Stellas	
Address:	Kleinschmidtstr. 48, 69115 Heidelberg, Germany	
Telephone:	+49 (0)6221 2 75 15	
E-mail:	pstellas@t-onlinde.de	
Date of birth:	3 April 1977	
Work experience		
1999 – present	Engaged on a part-time basis in the marketing office of a large IT company called BIBs. I participated in several major projects in the company.	
1997-1999	Employed in The Sport Shop in Walldorf on a part-time basis as a shop assistant.	
1996 to 1997	Employed as a sports trainer in Tarragona, Spain. I coached clients in tennis, squash, jogging and swimming. The work included giving advice on diet and supplements.	
Formal qualifications		
2003, March	Successfully completed my Staatsexamen at the University of Mannheim in Business Administration (equivalent to BSc Hons. or MSc).	
2001	Sat the Level 3 marketing exam of the London Chamber of Commerce which I passed with grade B.	
1997	Passed the Advanced Spanish exam of the Barcelona Chamber of Commerce with a distinction.	
1996	Passed my Abitur (approximately GCE A Level) with 1.3. (all subjects with A grades).	

Curriculum Vitae		
Other qualifications		
2001, March	Sat the spoken English exam for industry and commerce and passed with distinction	
2001, June	Achieved level 3 (advanced) of English for Business (EFB)	

3.1.6 Muster 2: Functional CV

Curriculum Vitae	
Employment	
March 1996 to present:	Sales Director, Floor Coverings Division, Viscose Corporation, London. Responsible for national sales team of 16 reps and overseas agents in 22 countries.
January 1993 – March 1996:	Sales Manager, Heritage Furnishings, Leeds. Responsible for national sales of all company products; led sales team of 7.
April 1991 – January 1993:	Senior Sales Representative, Heritage Furnishings.
May 1990 – April 1991:	Sales Representative, Heritage Furnishings.
September 1988 – May 1990:	Sales Trainee, Optimum Fabrics Ltd, Manchester.
Education St Stephen's College 1979 – 86; 8 O-levels, 2 A-levels Brighton Union Art College 1986 – 88; Design Diploma; Fabric Design Certificate.	
Career Profile My interest in design and fabrics led me into the furnishings trade where because of my technical and art training I was offered a trainee sales position. From then on point my career has grown to the point where I am now responsible for national and international sales of a range of innovative floor coverings (sales in 1997 £31.6 m); building the brands; marketing strategy and advertising agency liaison (3 accounts billing £2.8 m) – all figures on trade record. I now believe my experience and success have prepared me for a broader-based executive sales position, preferably with an international home products company.	
Personal Profile I have always been regarded as a good team worker which I believe has helped me to lead and inspire sales teams. I enjoy my work and am always prepared to put in extra hours a week of my own time. I am computer literate and write and design most of the company's sales literature. I am quite competitive and have won several sales awards during my career.	

Curriculum Vitae
Other I have a clean driver's licence. I have no problem relocating either within the UK or abroad. I am at present half-way through a part-time MBA course with the Open University.
Personal Age 31. Born 16/10/1978 British nationality Formerly married six years; now divorced; one 5-year old child (girl) Health: excellent Interests: tennis, patio gardening
Documentation Full sales history and detailed responsibilities can be supplied on request. References available on request.

3.2 Bewerbungsanschreiben richtig lesen

In a covering letter applicants are usually replying to an advert they seen. In their application they must demonstrate that they have understood the contents of the advert and are making a serious attempt at getting the position.

In einem Anschreiben antworten Jobanwärter:innen meistens auf eine vorher gesehene Stellenanzeige. In ihrer Bewerbung müssen sie deutlich machen, dass sie den Inhalt der Anzeige verstanden haben und ihre Motivation für die Position überzeugend darstellen.

- Is the purpose of your covering letter clear?
- Does the applicant's brief introduction in the covering letter arouse your interest?
- Is the applicant able to demonstrate his or her Unique Selling Point?
- Does the applicant ask for a reply?

- Wird der Zweck des Anschreibens deutlich?
- Weckt die Kurzvorstellung im Anschreiben Ihr Interesse?
- Gelingt es dem:der Bewerber:in ein Alleinstellungsmerkmal aufzuzeigen?
- Bittet der:die Bewerber:in um eine Rückmeldung?

3.2.1 Textbausteine: Anschreiben verstehen und analysieren

Der folgende Leitfaden zeigt, wie ein Anschreiben typischerweise aufgebaut ist, und hilft Ihnen mittels der Textbausteine beim Verständnis und der Analyse des Anschreibens.

Opening paragraphs	**Einleitung**
I am writing to you about the [*job title*] position advertised in the [*date*] edition of the [*newspaper*]. I am very interested in the position and my knowledge, skills and experience would be invaluable to your company.	Der Grund meines Schreibens ist die am [*Datum*] in der Ausgabe der [*Name der Zeitung*] ausgeschriebene Stelle. Die Position interessiert mich sehr und meine Kenntnisse, Fähigkeiten und Erfahrungen wären für Ihre Firma außerordentlich wertvoll.
[Name of contact], a close family friend, referred your name to me. I am graduating in December with a [name of degree]. I am interested in pursuing a career in [profession] and would appreciate the opportunity to interview with you for a full-time position.	[*Name der Kontaktperson*], ein enger Freund der Familie, hat mir Ihren Namen genannt. Im Dezember werde ich mein Studium mit einem [*Name des Hochschulabschlusses*] abschließen. Ich bin daran interessiert eine berufliche Laufbahn als [*Berufsbezeichnung*] einzuschlagen, und wäre über die Möglichkeit eines persönlichen Gesprächs über eine Vollzeitstelle dankbar.
I am a specialist in Human Resource Management. I am interested in [company name] because of its emphasis on teamwork and commitment to excellence. My knowledge, experience and drive to succeed would make me an excellent addition to your company.	Ich bin Spezialist:in für Personalwesen. Ich bin an [*Name der Firma*] interessiert, weil dort Teamarbeit und das Bekenntnis zur Leistung großgeschrieben werden. Meine Kenntnisse, Erfahrung und Wille zum Erfolg würden mich zu einer hervorragenden Ergänzung Ihrer Firma machen.
Main body paragraphs	**Hauptteil**
I believe that with your company's reputation and my fresh insight into [*specialist field*], I could make a significant contribution right away.	Ich glaube, dass ich mit dem Ruf Ihrer Firma und meinen neuen Kenntnissen in [*Spezialgebiet*] direkt einen wichtigen Beitrag leisten könnte.
With my background in [specialist field] I believe my unique skills would promote your company's continued growth and success.	Mit meinem Werdegang in [*Spezialgebiet*] glaube ich, dass meine einmaligen Fähigkeiten das kontinuierliche Wachstum und den Erfolg Ihrer Firma fördern würden.
I am someone who can work with customers. I believe I would make a noticeable difference at your company.	Ich bin jemand, der mit Kunden arbeiten kann. Ich glaube, ich würde in Ihrer Firma spürbar etwas bewegen.
Because I have XX years of experience in [field], I believe I fulfil your company's requirements.	Aufgrund meiner XX-jährigen Erfahrung im Bereich [*Fachgebiet*]glaube ich, den Bedürfnissen Ihrer Firma gerecht zu werden.
Since my background is in [field], I believe I could be a valuable contributor to your company's objectives.	Aufgrund meines Werdegangs in [*Fachgebiet*] glaube ich, dass ich einen wertvollen Beitrag leisten kann, um Ihre Unternehmensziele zu erreichen.

Closing paragraphs	Abschluss
Enclosed is a CV outlining my qualifications and achievements as well as recent personal publications. I look forward to meeting you in order to further discuss the position and my qualifications. Thank you for your time and consideration.	Beiliegend ein Lebenslauf, der sowohl meine Qualifikationen und Erfolge als auch meine persönlichen Veröffentlichungen kurz darstellt. Ich freue mich auf ein persönliches Treffen mit Ihnen, um darin die zu besetzende Stelle und meine Qualifikationen zu erörtern. Ich danke Ihnen für Ihre Zeit und Berücksichtigung.
I would appreciate the opportunity to meet with you to discuss my interest in working with your organization. I will contact your office in the next few days to see if it would be possible to arrange for an interview. In the meantime, please contact me if you would like any further information. Thank you for your time and consideration.	Ich würde mich über die Gelegenheit, mit Ihren über mein Interesse an einer Mitarbeit in Ihrer Firma sprechen zu können, freuen. Ich werde Ihr Büro in den nächsten Tagen kontaktieren, um einen möglichen Gesprächstermin zu vereinbaren. Wenn Sie zwischenzeitlich weitere Informationen benötigen, nehmen Sie bitte Kontakt mit mir auf.
I appreciate your consideration of my application and look forward to hearing from you soon.	Ich bedanke mich für die Berücksichtigung meiner Bewerbung und freue mich, bald wieder von Ihnen zu hören.
Thank you for your consideration. I look forward to your response.	Vielen Dank für Ihre Berücksichtigung. Ich freue mich auf Ihre Antwort.

3.2.2 Musteranschreiben 1: Management training programme

Dear Sir or Madam

I am responding to your advertisement in *The Independent* this morning regarding your management training programme.

I have recently completed a history degree at West London University and gained an upper second. I am aware that this is not a vocational course, but I consider the research disciplines acquired on the course to be highly applicable in the world of business.

I am an admirer of your products and have been for many years, and I would be very glad to be working in the environment in which they are produced.

I enclose my CV, which details my academic record and also gives information about the temporary work that I have undertaken.

I would be pleased to come for an interview whenever it is convenient to you.

Yours faithfully

3.2.3 Musteranschreiben 2: Junior clerk

Dear Mr XX

I am responding to the advertisement in today's Guardian for a junior clerk in your office.

I have just completed my education at Claverham College, where I passed 10 GCSEs, including English and mathematics. I am a confident person, have already worked in one office on a temporary basis, and have learned skills such as answering the telephone, so I feel I would be useful in your office. I am well organised and feel that the type of work described in your advertisement would be ideal for me.

I enclose my curriculum vitae, which gives more information about my education and experience. My schoolteachers will be able to provide references.

Yours faithfully

3.3 Vokabeln – Vocabulary input

Incorporating action words

In der folgenden Tabelle finden Sie »Action Words« (Aktivverben), mit denen Bewerber:innen ihre Leistung in den Vordergrund stellen. Nutzen Sie die Liste, um Bewerbungen auf solche Vokabeln hin zu analysieren.

English	German
Using initiative	**Initiative zeigen**
bought	gekauft
built	gebaut, aufgebaut
created	erstellt, geschaffen
designed	entworfen
engineered	entwickelt
forecast	vorhergesagt, prognostiziert
formed	gestaltet
generated	erstellt, generiert
improvised	improvisiert
initiated	initiiert

English	German
instigated	angeregt
launched	eingeführt
originated	hervorgebracht
pioneered	Pionierarbeit geleistet, Vorreiterrolle gespielt
Solving problems	**Probleme lösen**
advised	beraten
analysed	analysiert
combined	kombiniert
cut	eingespart, gekürzt
examined	untersucht
identified	identifiziert
investigated	untersucht
reduced	reduziert
reorganized	reorganisiert
resolved	geklärt, gelöst, beseitigt
revised	überarbeitet
solved	gelöst
streamlined	rationalisiert
trimmed	gekürzt
Managing situations	**Situationen bewältigen**
appointed	ernannt, berufen
appraised	bewertet
approved	genehmigt
controlled	kontrolliert
developed	entwickelt
directed	geführt, geleitet
drove	gelenkt, angetrieben
employed	eingesetzt, eingestellt
guided	geführt, geleitet
headed	angeführt, geleitet
inspired	inspiriert
led	geführt, geleitet

English	German
managed	geleitet, verwaltet
ran	betrieben, geleitet, geführt
Achieving targets	**Ziele erreichen**
accomplished	geschafft, fertig gestellt
achieved	erreicht
completed	erledigt, vollendet, fertig gestellt
conducted	durchgeführt, abgewickelt
delivered	gelieferte
demonstrated	aufgezeigt, demonstriert
exceeded	übertroffen
improved	verbessert
increased	gesteigert, erhöht
obtained	erhalten
produced	produziert, erzeugt
promoted	gefördert, vorangebracht
secured	sichergestellt, gesichert
selected	selektiert, ausgewählt

Nachfolgend finden Sie eine Auflistung aller wichtigen Fachbegriffe aus diesem Modul in alphabetischer Reihenfolge.

English	German
A level (= Advanced Level) *I took my A levels in 3 subjects.*	Abitur *Ich habe mein Abitur in 3 Fächern gemacht.*
academic record	akademische Bildung
accomplished	vollendet
admirer *I'm an admirer of Mozart's music.*	Bewunderer:in *Ich bin eine große Bewunderin der Musik von Mozart.*
appreciate something	etwas schätzen
apprenticeship	Lehre
approved *We are an approved examination centre of the London Chamber of Commerce.*	freigegeben, genehmigt, zugelassen *Wir sind ein zugelassenes Prüfungszentrum der Londoner Handelskammer.*

English	German
award a degree	einen akademischen Grad verleihen
Baccalaureate	Hochschulreife
certificate	Abschluss; Zertifikat
clean driving licence	Führerschein ohne Verkehrsregistereinträge
computer literate	über gute Computerkenntnisse verfügen
contributor	Mitarbeiter
convictions	Vorstrafen
covering letter	Anschreiben
curriculum vitae (CV)	Lebenslauf
degree *I have a degree from the University of Cambridge.*	Abschluss *Ich habe einen Abschluss von der Cambridge Universität.*
demonstrate	unter Beweis stellen
diploma	Diplom
distinction	Auszeichnung
DOB (date of birth)	Geburtstag
emphasise	betonen
exceed *The result exceeded my expectations.*	übertreffen *Das Ergebnis hat meine Erwartungen übertroffen.*
forecast *We had to revise our forecast for the next year.*	Vorhersage *Wir mussten unsere Vorhersage für das nächste Jahr revidieren.*
GCE/A level (= General Certificate of Education, Advanced Level)	Abitur
GCSE (= General Certificate of Secondary Education)	Mittlere Reife
get promotion	befördert werden
hands-on knowledge	Praxiswissen
head something *I headed the personnel department.*	etwas leiten *Ich leitete die Personalabteilung.*
improvise	improvisieren
insight	Einblick
instigate	anregen

English	German
intermediate knowledge	Mittelstufenkenntnisse
invaluable	äußerst wertvoll
job-hunting	auf Arbeitssuche sein
keep something in mind	etwas bedenken; berücksichtigen
liaison *I'm responsible for the liaison between head-quarters and subsidiary.*	Verbindung; Kontakt *Ich bin verantwortlich für den Kontakt zwischen Zentrale und Tochtergesellschaft.*
O level/GCSE (= General Certificate of Second-ary Education)	Mittlere Reife
one-page summary	1-seitige Zusammenfassung
outline	darlegen
part-time	Teilzeit
pioneer	Vorreiter sein
previous	vorherig
pursue a career in	einer Laufbahn nachgehen
recipient	Empfänger
relocation	Umzug
respond	antworten
restructure	umorganisieren
resume	Lebenslauf (USA)
revise	überarbeiten
sales director	Verkaufsleiter
streamline *We must streamline our administration.*	rationalisieren *Wir müssen unsere Verwaltung rationalisieren.*
testimonial *I was given a very good testimonial from my last employer.*	Arbeitszeugnis *Ich erhielt ein sehr gutes Arbeitszeugnis von meinem letzten Arbeitgeber.*
trainee	Kaufmännischer:e Auszubildender:e
training	Ausbildung
trim *I managed to trim the report down to 3 pages.*	kürzen *Ich konnte den Bericht auf 3 Seiten kürzen.*
vocational course	Ausbildungslehrgang

3.4 Übungen – Exercise time

Task 1: Fill in the gaps with the correct form of the word in brackets.

1. There were 200 ________________ at the conference in Cologne. (attend)
2. Germany certainly has more than 4 million ________________ people. (employ)
3. She ________________ top marks in her degree examinations. (achievement)
4. Anna hopes to pass her marketing test ________________. (success)
5. It's not ________________ the best way to always choose the cheapest offer. (necessary)

Task 2: Developing vocabulary. What's the opposite of:

1. employment = ________________
2. responsible = ________________
3. experience = ________________
4. literate = ________________
5. available = ________________
6. exclude = ________________
7. useful = ________________

Task 3: What are the terms for these definitions?

1. a letter which accompanies a CV = ________________
2. someone who gives advice on investment = ________________
3. something which is not permanent = ________________
4. person who is trying to get a job = ________________
5. somebody who writes a letter about the character and competence of an applicant = ________________

ARBEITSMITTEL AUF MYBOOK.HAUFE.DE

Auf der Internetseite zum Buch finden Sie weitere Übungen zu diesem Modul. Darüber hinaus stellen wir Ihnen exemplarische Lebensläufe und Bewerbungsanschreiben zur Verfügung:

- Muster eines US-amerikanischen Lebenslaufes
- Muster eines chronologisch strukturierten Lebenslaufes
- Muster eines Covering letter

4 Arbeitszeugnisse und Referenzen analysieren und schreiben

Zusammenfassung

In Ländern wie Großbritannien ist es durchaus üblich, Referenzen zu Bewerbern einzuholen. Wie Sie das machen und wie Sie selbst Referenzen und Arbeitszeugnisse schreiben, zeigen wir Ihnen in diesem Modul. Dazu bieten wir Ihnen

- *Textbausteine* und *Musterschreiben* für Referenzanfragen und Referenzschreiben,
- über 130 *Textbausteine* (u. a. zu 10 verschiedenen Soft Skills in verschiedenen Notenstufen), damit Sie rasch ein professionelles Zeugnis verfassen können, und
- eine *Checkliste* zur Gliederung eines Arbeitszeugnisses.

Spezifische *Vokabeln* finden Sie am Ende des Moduls sowie etliche *Übungen*, um Ihr neu erlerntes Wissen zu überprüfen. Die Lösungen finden Sie im Anhang dieses Buches.

Die Meilensteine: Von der Zeugnisanalyse bis zur Einladung zum Interview

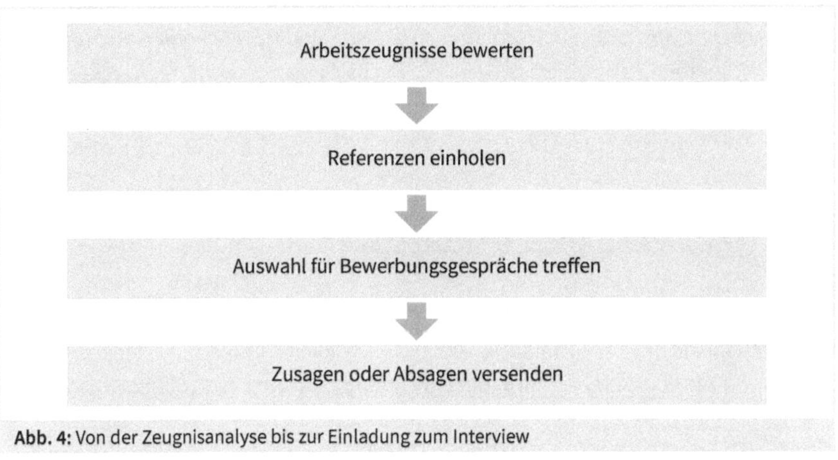

Abb. 4: Von der Zeugnisanalyse bis zur Einladung zum Interview

4.1 Geschäftliche Referenzen anfordern und Referenzen schreiben

Employers and recruitment professionals generally agree that references are important. These can verify that the candidate possesses the right skills for a job. A job application accompanied by several references will usually be taken seriously.

Arbeitgeber und Fachleute aus dem Bereich der Personalbeschaffung sind sich darüber einig, dass Referenzen wichtig sind. Diese können bestätigen, dass Kandidat:innen über die richtigen Fähigkeiten für die Arbeitsstelle verfügen. Eine mit Referenzen versehene Bewerbung wird üblicherweise ernst genommen.

When asking for a reference the referees should be given some guidance on what you want them to emphasise. References can be the deciding factor in a candidate getting the job or receiving a letter of rejection.
Primarily, a reference letter is used to introduce a person and vouch for his integrity, character and abilities.

Wenn Sie Referenzen einholen, sollten Sie den Referenzgebern erläutern, auf was es Ihnen in der Referenz ankommt. Referenzen können der entscheidende Faktor für eine Stellenzusage oder -absage sein.
Referenzen werden hauptsächlich dafür verwendet, eine Person vorzustellen und deren Integrität, Charakter und Fähigkeiten zu belegen.

4.1.1 Textbausteine: Asking for references

Sie möchten Referenzen über Bewerber:innen einholen. Im folgenden Leitfaden finden Sie nützliche Beispielformulierungen dazu – gegliedert nach Einleitung, dem Hintergrund für die Referenz, der Bitte um eine Referenz sowie einem Schlusssatz.

Opening	Einleitung
We are writing to you today because … Our company is currently looking for …	Wir schreiben Sie heute an, weil … Unsere Firma ist zurzeit auf der Suche nach …
The reason for this letter is to kindly ask you to …	Der Grund für dieses Schreiben ist, Sie höflich darum zu bitten, …
We were given your contact details by Ms XY who is currently applying for …	Wir erhielten Ihre Kontaktadresse durch Frau XY, welche sich zurzeit als … bewirbt.

Explaining the connection to the reference	Die Verbindung zur Referenz darlegen
Mr XY has informed us that you were willing to write a reference for Ms YY	Herr XY hat uns darüber informiert, dass Sie bereit sind, für Frau YY eine Referenz zu schreiben.
We learned from Mr XY that he was working from … to … in your company. We would be very happy if you …	Wir erfuhren von Herrn XY, dass er in der Zeit von … bis … für Ihre Firma gearbeitet hat. Wir würden uns darüber freuen, wenn Sie …
Mr XY has applied to us for the position of …, we therefore would therefore like to ask you to …	Mr XY hat sich bei uns für die Stelle als … beworben, daher möchten wir Sie bitten, …
Asking for the reference	**Um eine Referenz bitten**
I trust you feel able to write a reference for	Ich hoffe, dass Sie eine Referenz für … schreiben können.
I'd like to ask you for a reference for …	Ich würde Sie gerne um eine Referenz für … bitten.
I hope you will be happy to write a reference for …	Ich hoffe, es macht Ihnen nichts aus, eine Referenz für … zu schreiben.
Would you be kind enough to supply us with a reference for …	Wären Sie so nett und würden uns eine Referenz für … zukommen lassen.
We'd be grateful if you could write a reference for … .	Wir wären Ihnen sehr dankbar, wenn Sie eine Referenz für … schreiben könnten.
Closing	**Abschluss**
I'd like to thank you for writing this reference.	Ich möchte mich bei Ihnen für die Erstellung dieser Referenz bedanken.
I look forward to receiving your letter of reference.	Ich freue mich, die Referenz zu erhalten.
I thank you in advance for …	Ich danke Ihnen im Voraus für …
I hope you will feel able to send a reference.	Ich hoffe, Sie fühlen sich in der Lage eine Referenz zu schicken.
I'd like to take the opportunity of thanking you for writing this reference.	Ich möchte die Gelegenheit wahrnehmen, Ihnen für die Erstellung dieser Referenz zu danken.
Thank you for your time and consideration in this matter.	Vielen Dank für Ihre Zeit und Mühe in dieser Angelegenheit.

4.1.2 Musterschreiben: Asking for a reference

Beispiel 1: Asking for a reference
Dear Mr XY

We were given your contact details by Ms XY who is presently applying for the position of sales manager in our company.

Ms XY informed us that she has been working for you since February 2003. We learned that she has been responsible for liaison with customer contacts for the last 4 years. We believe Ms XY is a suitable candidate for the vacant position in our company and would be very happy if you could provide us with a written reference letter outlining her duties, skills and qualifications. In addition could you also let us know whether you believe that Ms XY would be suitable for the position.

I look forward to receiving your letter of reference.

Best regards

Beispiel 2: Asking for a reference
Dear Mr XY

The purpose of this letter is to ask you for a reference for Mr XY who has given us your name as referee. Please support us in trying to evaluate whether he would fit the position by completing the attached reference form. The information that you provide us with will be kept confidential.

Thank you for your time and consideration in this matter.

Best regards

4.1.3 Textbausteine: Writing a reference letter

Nachfolgend finden Sie Musterformulierungen für die Erstellung von Referenzen. Nach dem Leitfaden folgen zwei Beispiele.

Salutation	Anrede
To whom it may concern	An die zuständige Abteilung/Stelle
Dear Mr; Mrs; Ms [name of employer]	Sehr geehrter Herr … / Sehr geehrte Frau … [Name des Arbeitgebers]

Opening	**Einleitung**
I am writing this reference letter at the request of Ms XY who is applying for the position of [job title] with your company.	Ich schreibe diese Referenz auf Wunsch von Frau XY, die sich für die Stelle als [Bezeichnung der Stelle] in Ihrer Firma bewirbt.
I am writing this letter at the request of Dr XY.	Ich schreibe diesen Brief auf Wunsch von Herrn Dr. XY.
This is in response to your recent request for a letter of reference for Mr XY.	Ich nehme Bezug auf Ihre kürzliche Anforderung einer Referenz für Herrn XY.
I am pleased to be able to write this letter of reference for Mr XY.	Ich freue mich, diese Referenz für Herrn XY schreiben zu können.
Making an assessment	**Eine Einschätzung abgeben**
In my opinion, Dr XY is a hard-working self-starter who invariably understands exactly what a project is all about.	Meiner Meinung nach ist Dr. XY ein hart arbeitender Mensch mit viel Eigeninitiative, der stets genau versteht, um was es bei einem Projekt geht.
Ms XY consistently produces high quality work in a timely fashion.	Frau XY erbringt beständig und termintreu qualitativ hochwertige Arbeitsleistungen.
The only area of weakness that I ever noted in [name]'s performance was …	Den einzigen Schwachpunkt, den ich jemals in der Arbeitsleistung von (Name) wahrnahm war …
He is a resourceful, creative and solution-oriented individual.	Er ist ein einfallsreicher, kreativer und lösungsorientierter Mensch.
She knows how to get things done quickly and effectively.	Sie kann Dinge schnell und effektiv erledigen.
Closing	**Abschluss**
I am therefore very pleased to be able to recommend Ms XY for this position.	Ich bin daher sehr froh, Frau XY für diese Position empfehlen zu konnen.
Based on my personal experience when working with Ms XY, I can recommend her very highly.	Aufgrund meiner persönlichen Erfahrung in der Zusammenarbeit mit Frau XY kann ich sie in hohem Maße empfehlen.
I would like to recommend Ms XY for the job of … in your organisation.	Gerne empfehle ich Frau XY für die Stelle als … in Ihrem Betrieb.
In conclusion, I would highly recommend Dr. XY for your company.	Abschließend würde ich Herrn Dr. XY für Ihr Unternehmen in hohem Maße empfehlen.

4.1.4 Musterschreiben: Letter of reference

Beispiel 1: Letter of reference
To Whom it May Concern:

I highly recommend Ms Smith as a candidate for employment. Ms Smith was employed in our company as a clerk from 2005 – 2008. She was responsible for office support including word processing, scheduling appointments and creating brochures, newsletters and other office documentation.

Ms Smith has excellent communication skills. In addition, she is extremely well-organized, reliable and computer literate. She can work independently and is able to ensure that the job gets done. She is flexible and willing to work on any project that is delegated to her. Ms Smith was quick to volunteer to assist in other areas of company operations.

The employee would be a great asset for your company. If you have any further questions with regard to her background or qualifications, please do not hesitate to contact me.

Yours faithfully

Beispiel 2: Letter of reference
Dear Mr XY

Thank you for your letter of 2 May regarding the employment of Mr Roger Oldham.

My CEO has asked me to inform you that Mr Oldham has worked here as cashier for a period of three years. During this time he has proved himself to be efficient and capable, and his only reason for wishing to leave us is to seek a post with potential for promotion. Unfortunately, we are unable to pay him a higher salary, and there is no possibility of promotion at this time in the company.

If he is successful in his application to you, our company will be sorry to lose his services. At the same time, we would wish him well in his new position.

Yours sincerely

4.2 Ein Arbeitszeugnis verfassen

A testimonial is issued at the request of the employee at the time of termination of the employer-employee relationship. The employee can choose between a simple and a detailed testimonial. The simple testimonial includes only dates, duration and type of employment. In contrast, the detailed testimonial will give statements about performance, social behaviour and qualification.

It is used for applying for a new job, change of superior, restructuring within the company, personal changes and training leave entitlement.

Ein Arbeitszeugnis wird auf Wunsch des Arbeitnehmers meistens zum Beendigungszeitpunkt des Arbeitsverhältnisses erstellt. Der Arbeitnehmer kann zwischen einem einfachen und einem qualifizierten Zeugnis wählen. Das einfache Zeugnis beinhaltet nur die persönlichen Daten und die Dauer und den Typ der Beschäftigung. Dagegen macht das detaillierte Zeugnis Aussagen über die Leistungsfähigkeit, das Sozialverhalten und die Qualifikation.
Es wird für die Bewerbung für einen neuen Arbeitsplatz, den Wechsel von Vorgesetzten, innerbetriebliche Restrukturierungen, persönliche Veränderungen und bei Anspruch auf Bildungsurlaub verwendet.

4.2.1 Checkliste: Aufbau eines Arbeitszeugnisses

Wie Sie ein Arbeitszeugnis aufbauen sollten, haben wir in der folgenden Übersicht für Sie zusammengestellt.

Section	Content	[x]
Title	Work reference/Intermediate testimonial/Educational reference/Trainee testimonial	
Introductory phrasing	• Personal details: full name of employee, date and place of birth • Duration of relationship • Job title • Location within company	
Description of activities	• Currently performed tasks • Position in hierarchy • Competences • Responsibilities	

Section	Content	[x]
Evaluation of performance	• Expertise and know-how *(Kriterium 1)* • Special skills *(Kriterium 2)* • Perception and problem solving *(Kriterium 3)* • Ability to cope with pressure *(Kriterium 4)* • Approach to work *(Kriterium 5)* • Reliability *(Kriterium 6)* • Work result *(Kriterium 7)* • Leadership qualities *(Kriterium 8)* (nur bei Führungskräften) • Soft skills *(Kriterium 9)* • Summary of performance *(Kriterium 10)*	
Social evaluation	• Behaviour towards superiors/colleagues/third parties (customers, suppliers, business associates)	
Final paragraph	• Reason for issue of testimonial • Thank-you and statement of regret	
Signature	• Place and date of issue • Signature of author and his/her position	

4.2.2 Textbausteine zu allen wichtigen Kriterien und in drei Bewertungsstufen

Auch im Englischen gibt es eine Zeugnissprache. Sowohl für die Austeller als auch die Empfänger eines Zeugnisses, ist es wichtig, diese Klauseln zu kennen. Im Folgenden finden Sie praxiserprobte Textbausteine, die Sie zum Erstellen eines Zeugsnisses verwenden können bzw. die Ihnen beim Lesen des Arbeitszeugnisses Ihres Bewerbers behilflich sind.

Einleitung – Introductory phrasing

It is my pleasure to confirm that Ms XY has worked as … in our company.	Gerne bestätige ich, dass Frau XY als … in unserem Unternehmen gearbeitet hat.
I can confirm at this point that Mr XY …	Ich kann an dieser Stelle bestätigen, dass Herr XY …
Mrs XY, born on …, worked from … to … in our company.	Frau XY, geboren am … arbeitete von … bis … in unserem Unternehmen.
Dr. XY, working at present in the Sales department, has been an employee of ours for … years	Dr. XY, der zurzeit im Verkauf arbeitet, ist seit … Jahren Betriebsangehöriger.
Mr XY is employed as XX.	Herr XY ist als XX beschäftigt.

Tätigkeitsbeschreibung – Description of activities

Mr XY has been performing the task/job of … for … years.	Herr XY arbeitet seit … Jahren als …
She/he has developed and implemented a system in …	Sie/Er entwickelte und implementierte ein System für …
Dr XY is responsible for six employees in the department of …	Dr. XY ist in der Abteilung … verantwortlich für sechs Angestellte.
Mr XY's duties include …	Die Aufgaben von Herrn XY umfassen …
Mrs XY is head of department for …	Frau XY ist Leiterin der Abteilung …

Kriterium 1: Fachwissen und Fachkenntnis – Expertise and know-how

Excellent	Good	Average
Mr [Name] possesses exceptionally profound expertise, which he consistently used effectively and successfully in his work. He was able to impart this expertise to his employees without reservation.	Mr [Name] possesses extensive expertise, which he used effectively and successfully in his work. He was able to impart this expertise to his employees.	Mr [Name] possesses quite extensive expertise, which he used successfully in his work. He was able to impart this expertise to his employees.
Mr [Name] continuously impressed us with his excellent expertise, even in tangential areas, which he always used confidently and competently in his work.	Mr [Name] impressed us with very good expertise, which he also used confidently and competently in his work.	Mr [Name] was recognised for his expertise and his calm and confident way of applying his knowledge in his work.
Mr [Name] always used his expertise, which was of extraordinary depth and breadth, confidently and very efficiently in his daily tasks.	Mr [Name] used his impressive expertise confidently and efficiently in his daily tasks.	Mr [Name] used his practical expertise confidently in his daily work.

Kriterium 2: Besondere Fähigkeiten – Special skills

Excellent	Good	Average
We are pleased to confirm her extraordinary economic competence. Her outstanding entrepreneurial and strategic thinking and actions earned her the utmost respect from management and employees.	We are pleased to confirm her high level of economic competence. Her well developed entrepreneurial and strategic thinking and actions earned her the greatest respect from management and employees.	We are pleased to confirm her economic competence. Her entrepreneurial and strategic thinking and actions earned her great respect from management and employees.

Excellent	Good	Average
His excellent language skills in … deserve a special mention. He was always a valued dialogue partner for our international customers and business partners.	His very good language skills in … deserve a special mention. He was always a valued dialogue partner for our international customers and business partners.	His good language skills in … deserve a special mention. He was always a valued contact person for our international customers and business partners.
She possesses outstanding IT expertise. She was always a valued resource, both on the user side (office products, business solutions) and in the acquisition of new hard and software.	She possesses very good IT knowledge. She was always a valued resource, both on the user side (office products, business solutions) and in the acquisition of new hard and software.	She possesses good IT knowledge. She was a valued resource both on the user side (office products, business solutions) and in the acquisition of new hard and software.

Kriterium 3: Auffassungsgabe/Problemlösekompetenz – Perception and problem solving

Excellent	Good	Average
Due to his precise analytical abilities and his speed of perception he found outstanding solutions which he consistently and successfully put into practice.	Mr [Name] is a dedicated manager who always successfully accomplished his assignments with complete commitment.	Mr [Name] is a dedicated manager who successfully accomplished his assignments with complete commitment.
Due to her extreme speed of perception and her methodical approach he quickly found an intelligent and elegant solution even to difficult problems. She always had a firm grasp of all tasks in her area of responsibility.	Ms [Name] accomplished her assignments with great dedication and personal commitment during her entire employment with our company.	Ms [Name] accomplished her assignments with average dedication and personal commitment during her entire employment with our company.
Mr [Name]'s instant comprehension enabled him to grasp new developments immediately and precisely assess their consequences.	Mr [Name] takes the initiative himself and shows a great level of commitment to both our company and our customers.	Mr [Name] takes the initiative and demonstrates a high level of commitment to both our company and our customers.

Kriterium 4: Belastbarkeit – Ability to cope with pressure

Excellent	Good	Average
Even in stressful situations he achieved excellent performance levels, both qualitatively and quantitatively, and could cope under the heaviest of workloads.	Even under great pressure he kept control of the situation, acted thoughtfully and accomplished all tasks well.	He was able to manage heavy workloads.
She was an exceptionally resilient employee, who mastered the high demands of her important position very well even under difficult circumstances and intense time pressure.	She was always a resilient employee, and the quality of her work remained consistently good despite varying challenges.	Even under difficult working conditions and great pressure, he always met our expectations to a satisfactory degree.
Even under the most difficult of working conditions, he accomplished his tasks in an outstanding manner.	Even under difficult working conditions, he accomplished his tasks well.	Even under great pressure he kept control of the situation, acted thoughtfully and accomplished all tasks in a satisfactory manner.

Kriterium 5: Arbeitsweise – Approach to work

Excellent	Good	Average
Mr [Name] always worked extremely conscientiously and precisely.	Mr [Name] always worked very, conscientiously and precisely.	Mr [Name] always worked carefully, conscientiously and precisely.
Ms [Name]'s approach to her work was always very well planned, extremely timely and result oriented.	Ms [Name]'s approach to her work was always well planned, timely and result oriented.	Ms [Name]'s approach to her work was always well planned, systematic and result oriented.
Mr [Name] always worked in an extremely timely, result oriented and precise manner.	Mr [Name] always worked in a very timely, yet extremely result oriented and precise manner.	Mr [Name] always worked in a timely, yet result oriented and precise manner.

Kriterium 6: Zuverlässigkeit – Reliability

Excellent	Good	Average
Her working methods were always marked by trustworthiness and absolute dependability.	Her working methods were marked by trustworthiness and dependability.	Her worked dependably and accurately.
She always stood out due to his extraordinary dependability.	She always worked dependably and very accurately.	She impressed us with his high level of dependability.
She always impressed us with his high level of dependability	She generally worked dependably and very accurately.	She was generally reliable.

Kriterium 7: Arbeitsergebnis – Work result

Excellent	Good	Average
Even for the most difficult challenges she found and implemented very effective solutions and always achieved excellent results.	Even for difficult challenges she found and implemented very effective solutions and always achieved good results.	For difficult challenges Ms [Name] found and implemented appropirate solutions and always achieved satisfactory results.
Even for the most difficult challenges, Mr [Name] found very effective solutions, which he always implemented successfully, achieving very good results.	Even for difficult challenges, Mr [Name] found very effective solutions, which he implemented successfully, always achieving good results.	For difficult challenges, Mr [Name] found effective solutions, which he implemented successfully, always achieving very solid results.
Ms [Name] achieved excellent results in all situations.	Ms [Name] achieved good results in all situations.	Ms [Name] achieved satisfactory results in all situations.

Kriterium 8: Führungskompetenz – Leadership qualities

Excellent	Good	Average
Mr [Name] gained the confidence of his employees and motivated them through cooperative leadership. He was always fully accepted as a superior, and his team not only met but often exceeded our high expectations.	He enjoyed the recognition and appreciation of his employees, as he assigned them tasks in keeping with their abilities and achieved good results with his team.	His employees performed well above average under his goal oriented leadership.

Excellent	Good	Average
Due to her cooperative management style, Ms [Name] was always well accepted and well respected by her employees, who invariably performed exceptionally well under her leadership.	Her team always achieved good results under her relevant and individual instruction. Her leadership qualities met with the complete acceptance of her superiors as well as her employees.	She was always able to motivate her employees and to lead them to achieve very solid results.
Mr [Name] gained the confidence of his employees and motivated them through cooperative leadership. He was always fully accepted as a superior, and his team always met our high expectations.	He enjoyed the confidence of his employees and fostered cooperation. Mr [Name] kept his team informed, encouraged further education, delegated tasks and responsibilities sensibly and thus achieved good results.	He practised a cooperative style of leadership and with his team he achieved very satisfactory results.

Kriterium 9: Soft Skills – Soft skills

Excellent	Good	Average
As a manager, Ms [Name] continually demonstrated her excellent integration abilities. She always excelled at including the members of her team in the decision-making process according to their personalities and competence, thereby creating an outstanding working atmosphere in her team.	As a manager, Ms. [Name] continually demonstrated her good integration abilities. She always succeeded in including all members of her team in the decision-making process according to their personalities and competence, thereby creating a very good working atmosphere in her team.	As a manager, Ms [Name] continually demonstrated her good integration abilities. She succeeded at including all members of her team in the decision-making process according to their personalities and competence, thereby creating a positive working atmosphere in her team.
Ms [Name] was always capable of managing conflicts. Her exceptionally constructive behaviour, thoughtful actions and respect for her associates allowed her to create a very positive working atmosphere in her team.	Ms [Name] was always capable of managing conflicts. Her constructive behaviour, thoughtful actions and respect for her associates allowed her to create a very positive working atmosphere in her team.	Ms [Name] was capable of managing conflicts. Her constructive behaviour and thoughtful actions allowed her to create a good working atmosphere in her team.

Excellent	Good	Average
As a manager, Ms [Name] was continually and in every respect open to feedback and suggestions for improvement from her employees. She always viewed criticism as a chance for improvement and personal growth. In return, she was always able to give positive as well as negative feedback to her employees at the appropriate times.	As a manager, Ms [Name] was continually open to feedback and suggestions for improvement from her employees. She always viewed criticism as a chance for improvement and personal growth. In return, she was always able to give positive and negative feedback to her employees at the appropriate times.	As a manager, Ms [Name] was open to feedback and suggestions for improvement from her employees. She always viewed criticism as a chance for improvement and personal growth. In return, she was able to give positive as well as negative feedback to her employees at the appropriate times.

Kriterium 10: Zusammenfassende Leistungsbeurteilung – Summary of performance

Excellent	Good	Average
Mr [Name] always completed tasks assigned to him to our utmost satisfaction.	Mr [Name] mostly completed tasks assigned to him to our complete satisfaction.	Mr [Name] mostly managed tasks assigned to him to our satisfaction.
Mr [Name] met and often exceeded our very high expectations.	We were always very satisfied with Mr [Name]'s performance.	Mr [Name] met our expectations in the best possible manner.
Mr [Name] always completed tasks assigned to him to our utmost satisfaction.	Mr [Name] always completed tasks assigned to him to our complete satisfaction.	Mr [Name]'s performance was entirely satisfactory.

Verhaltensbeurteilung – Social evaluation

Excellent	Good	Average
Ms [Name] got on very well with all associates and always communicated with them in his friendly, open and accommodating way. Her behaviour towards superiors, colleagues and external associates was always exemplary.	Ms [Name] got on well with all associates and always communicated with them in her friendly, open and accommodating way. Her behaviour towards superiors, colleagues and customers was always flawless.	Ms. [Name] was esteemed by all associates because of her friendly and conciliatory personality.

Excellent	Good	Average
Ms [Name] was highly esteemed by all associates because of her friendly and balanced personality, as she always fostered good cooperation and a team atmosphere. Her behaviour was always exemplary.	Ms [Name] was highly esteemed by all associates because of her friendly and balanced personality, as she always fostered good cooperation and team atmosphere. Her behaviour towards superiors, colleagues and customers was always flawless.	Ms [Name] was esteemed by all associates because of her friendly and balanced personality, as she actively fostered good cooperation and team atmosphere.
Ms [Name]'s behaviour towards superiors, colleagues and customers was always exemplary and loyal.	Ms [Name]'s behaviour was exemplary. She was greatly esteemed by her superiors, colleagues and customers.	Ms [Name] impressed us professionally and personally, and earned the recognition of her superiors, colleagues and customers.

Schlussformulierung – Final paragraph

Excellent	Good	Average
We thank Mr [Name] for the many years of constant cooperation and very much regret his leaving our company. We wish this exemplary colleague all the best, lots of luck and success in both his professional and personal life.	We truly regret his decision, thank him for his valuable service, and wish him all the best and continued success for his professional as well as personal future.	We regret his leaving, as it means that we are losing a professional associate. We can recommend him as a good employee and thank him for his first rate performance. For his onward professional and personal journey we wish him all the best and lots of success.
We regret her leaving very much, as it means that we are losing a valued professional associate. We can wholeheartedly recommend her as an employee. We thank her for her continually outstanding performance. For her onward professional and personal journey we wish her all the best and continued success.	We truly regret to see her go. We want to be sure to extend a special thanks for her work in our company. We wish her all the best and well deserved success for both her professional and personal future.	We regret her decision, thank her for her cooperation, and wish her all the best and continued success for the future.

4.2.3 Musterschreiben: Arbeitszeugnis – Testimonal

Beispiel 1: Testimonal
To Whom it May Concern:

It is my pleasure to confirm that Ms Paula Langdon has been employed by Global Trading since 12th March, 2007. Paula's main duties consist of supporting our distributors and clients with any technical problems they might experience on our testing equipment, on-site technical visits, and training of distributors/staff throughout the region, as well as organising and running our national customer support centre. This task includes the recruitment of technical staff.

As a leading project manager, Paula designed and implemented a regional online customer support help desk system that has been working perfectly, therefore enabling us to streamline our customer support. This has helped us to considerably increase our customers' confidence in our products and our team.

Throughout her time with our company, Paula has always been a responsible employee, offering a great amount of energy, initiative and responsibility in all aspects of the job. Her analytical abilities and her hands-on approach have been vital to the smooth running of our daily business. Thanks to her effective communication skills and friendly nature, Paula is well respected by her superiors and colleagues alike and we appreciate her as someone who can be relied upon in any situation. Her know-how is far above average and she is extremely thorough.

Paula's special qualification is her language skills. She is fluent in German, Russian and French, and is thus able to understand our customers needs and attitudes. Paula's results and her personal behaviour have always been exemplary.

It is with great pleasure that I provide this reference in recognition of Paula's good work. I do not hesitate to recommend Paula for any position where an outstanding customer service manager and a skilful team player is needed. We thank Paula for her loyalty and excellent performance.

Please feel free to get in touch with me personally if you have any further questions.

Nottingham, 2 May 2010

Mark Clerks

(CEO)

Beispiel 2: Testimonal (short)
To whom it may concern:

I confirm that Dr Paul Petersen was employed as Procurement Manager with this organisation from 1 January 2003 to 31 December 2009 and was/is paid (salary, plus bonus and benefits as applicable).

His job involved responsibility for purchasing raw materials as well as coordinating the random sampling. Peter is a skilled negotiator and is fluent in English and Russian. In addition he is a good communicator and a reliable colleague.

I would be happy to re-employ Peter as I consider him to be a valuable member of the team, who always achieved good results and fulfilled all expectations.

Yours faithfully

4.3 Vokabeln – Vocabulary input

In dieser Liste finden Sie alle wichtigen Fachbegriffe und viele andere nützliche Wörter rund um das Thema Referenzen und Arbeitszeugnisse.

English	German
ability *She has the ability to manage the department.*	Fähigkeit; Können; in der Lage sein *Sie ist in der Lage, die Abteilung zu leiten.*
above average	überdurchschnittlich
analytical abilities	analytische Fähigkeiten
asset for your company	Gewinn für Ihr Unternehmen
capable *She is very capable.*	fähig *Sie ist sehr kompetent.*
carefully	sorgfältig
commitment; personal ~	persönlicher Einsatz
communication skills	kommunikative Fähigkeiten
confidential; keep ~	vertraulich behandeln
confirm	bestätigen
contact details	Kontaktdaten
customer‘s confidence	Kundenvertrauen

English	German
decision-making process	Entscheidungsprozess
deserve a special mention *I think his achievements deserve a special mention.*	eine besondere Erwähnung verdienen *Ich glaube, seine Erfolge verdienen eine besondere Erwähnung.*
detailed testimonial	qualifiziertes Zeugnis
distributors	Händler
duties	Aufgaben
educational leave	Bildungsurlaub
employer-employee relationship	Arbeitsverhältnis
entitlement *Your holiday entitlement is 30 days.*	Anspruch *Ihr Urlaubsanspruch beträgt 30 Tage.*
entrepreneurial and strategic thinking	unternehmerisches und strategisches Denken
esteemed *My esteemed colleague is leaving us today.*	angesehen; geschätzt *Mein geschätzter Kollege verlässt uns heute.*
evaluate	bewerten
exceeded our high expectations	unsere hohen Erwartungen übertroffen
foster	unterstützen
further education	Weiterbildung
grasp *Accountants have no grasp of technology.*	Verständnis *Buchhaltern fehlt das Verständnis für Technologie.*
head of department	Abteilungsleiter
impart	mitteilen
implement	umsetzen, anwenden
integration abilities	Fähigkeiten sich zu integrieren
leadership performance	Führungsleistung
liaison	Zusammenarbeit
negotiator	Verhandlungsführer
on-site *The service technician checks the machine on-site.*	vor Ort *Der Servicetechniker überprüft die Maschine vor Ort.*
outline	erläutern
outstanding	außerordentlich

English	German
perception	Auffassungsgabe
performance	Leistung
project leader	Projektleiter
recommend *We are happy to recommend her for this job.*	empfehlen *Gerne empfehlen wir sie für diese Position.*
references	Referenzen
reliable	verlässlich
resilient *One of his plus-points is that he's very resilient.*	belastbar *Einer seiner Pluspunkte ist, dass er sehr belastbar ist.*
result-oriented	ergebnisorientiert
skill	Fähigkeit
specialist knowledge	Fachkenntnisse
suitable candidate	Geeigneter:e Bewerber:in
testimonial	Arbeitszeugnis
trustworthiness	Vertrauenswürdigkeit
utmost respect	höchste Anerkenntnis
vacant position	offene Stelle
verify	bestätigen

4.4 Übungen – Exercise time

Task 1: What is the term that best fits the following definitions?

1. Writing (or a telephone call) that gives you information about an applicant = t________________ or r________________
2. The opening of a business letter beginning »Dear ...« = s________________
3. The results you achieve in your work = p________________
4. Special information or knowledge you have of your work = k________________
5. Special salutation for an open testimonial = t________________ w________________ i________________ m________________ c________________

Task 2: Complete the sentences with the appropriate translation of the German word in brackets.

1. The candidate for this position must prove his ________________ (*Zuverlässigkeit*).
2. You don't get far nowadays if you neglect your ________________ (*Weiterbildung*).
3. Our company will indeed be ________________ (*bedauern*) to lose her services.
4. I would like to ________________ (*Bezug nehmen auf*) to several points in your letter of application.
5. Employees will be made ________________ (*entlassen*) due to the lack of orders.

Task 3: Find words which are the opposite of the following.

1. personal = ________________
2. include = ________________
3. satisfaction = ________________
4. applicable = ________________
5. future = ________________
6. social = ________________
7. attractive = ________________

ARBEITSMITTEL AUF MYBOOK.HAUFE.DE

Auf der Internetseite zum Buch finden Sie weitere Übungen und praktische Arbeitsmittel zu diesem Modul:

- Wir bieten Ihnen weitere Musterschreiben, um Referenzen einzuholen, sowie Muster für Referenzschreiben. Diese Vorlagen können Sie individuell anpassen, ausdrucken, speichern oder per E-Mail versenden.
- Mit einer Checkliste können Sie sowohl schriftlich als auch telefonisch eingegangene Referenzen prüfen. Praxiserprobte Bewertungsfragen helfen Ihnen dabei.
- Darüber hinaus finden Sie zwei Arbeitszeugnismuster: Eines für die Position eines Verkaufsmanagers sowie ein allgemeines Arbeitszeugnismuster, dass Sie individuell an die Daten des scheidenden Mitarbeiters anpassen.

5 Bewerber:innen interviewen und bewerten

Zusammenfassung

Ein Bewerberinterview nicht in der eigenen Muttersprache zu führen, ist eine große Herausforderung. Damit Sie sicher und erfolgreich solche Interviews führen können, unterstützen wir Sie mit zahlreichen Arbeitsmitteln:

- Ein umfangreicher *Interviewleitfaden* gibt Ihnen für ein strukturiertes Bewerbergespräch die richtigen Fragen an die Hand.
- Die Übersicht *Fünf Phasen eines Bewerberinterviews* zeigt Ihnen, wie Sie das Gespräch sinnvoll gliedern und welche Themen Sie jeweils ansprechen sollten.
- Beispieldialoge helfen Ihnen dabei, sich in die Situation einzufinden, Bewerber:innen zu interviewen.

Bereiten Sie sich mit der themenspezifischen *Vokabelliste* auf das Interview vor. Und testen Sie mit den *Übungen* Ihre Kenntnisse. Welche weiteren *Arbeitsmittel* wir für Sie auf mybook.haufe.de bereithalten, haben wir am Ende dieses Moduls notiert.

Die Meilensteine: Bewerberinterviews erfolgreich durchführen

Abb. 5: Bewerberinterviews erfolgreich durchführen

5.1 Bewerber:innen professionell interviewen

An interview is an oral questionnaire which tries to predict future job performance on the basis of the candidate's answers.
This tool is very widely used to select the best candidate for a position.

There are different ways this can be done. One efficient and common approach is to conduct a structured interview.
The characteristic of structured interviews is that you will ask several applicants the same question. This has the big advantage that the results of the interviews are easier to compare.

Ziel des Bewerberinterviews ist es, die zukünftigen Arbeitsleistungen des Bewerbers bzw. der Bewerberin aufgrund seiner/ihrer Antworten einzuschätzen. Dieses Instrument wird sehr häufig benutzt, um die optimale Besetzung für eine Position zu identifiziern.
Es gibt verschiedene Möglichkeiten, wie Sie dabei vorgehen können. Eine effiziente und gebräuchliche Art ist, ein strukturiertes Interview zu führen.
Das Kennzeichen von strukturierten Interviews besteht darin, dass Sie mehreren Bewerbern weitgehend die gleichen Fragen stellen. Das hat für Sie den großen Vorteil, dass Sie die Ergebnisse der Gespräche untereinander besser vergleichen können.

5.1.1 Fünf Phasen eines Bewerberinterviews

We recommend structuring the interview into five phases:
Phase one: You start with a warm-up consisting of a welcome, introduction of participants and a short presentation of the company.

In the **second phase** the applicant will have the chance to present himself/herself. He or she can do this on the basis of his or her CV.
In the **third phase** you may ask questions about the applicant's CV and enquire about his motivation.

Wir empfehlen Ihnen, das Bewerberinterview in fünf Phasen zu gliedern:
Erste Phase: Sie beginnen mit einem Warm-up, das aus der Begrüßung, der gegenseitigen Vorstellung und einer kurzen Präsentation des Unternehmens besteht.
In der **zweiten Phase** geben Sie dem:der Bewerber:in die Möglichkeit, sich selbst zu präsentieren. Dies kann er anhand seines Lebenslaufes tun.
In der **dritten Phase** stellen Sie Rückfragen zum Lebenslauf und hinterfragen die Motivation und Beweggründe für die Motivation.

In **phase four** you ask the applicant about his or her skills and experience. As a guideline you should use the requirement profile in which you have fixed the competencies required for the vacant position. Don't just ask simple questions but ask further questions that do dig deeper into the subject.

In der **vierten Phase** stellen Sie Fragen zu Fähigkeiten und Erfahrungen. Nutzen Sie dazu als Leitfaden das Anforderungsprofil, in dem Sie die für die zu besetzende Stelle erforderlichen Kompetenzen fixiert haben. Belassen Sie es nicht bei einfachen Fragen, sondern haken Sie immer wieder nach und stellen Sie Vertiefungsfragen.

In the **fifth phase** you conclude the interview, clarify any open questions and discuss the further procedure.

In der **fünften Phase** beenden Sie das Gespräch, klären offene Fragen und besprechen das weitere Vorgehen.

5.1.2 Gesprächsleitfaden: Strukturiertes Bewerberinterview in fünf Phasen

Phase 1: Warm-up

Formal introductions	**Begrüßung**
Formal first meeting: You: How do you do, Mr XY? Reply: How do you do? (accompanied by handshaking) You: Pleased/Glad to meet you. Reply: Likewise./Pleased to meet you, too.	Erstes förmliches Treffen: Sie: Wie geht es Ihnen Herr XY? Antwort: Wie geht es Ihnen? (begleitet von einem Händedruck) Sie: Schön Sie zu sehen/treffen. Antwort: Das Vergnügen ist ganz meinerseits.
At a second, more informal meeting: You: How are you?/How are you doing? Reply: Fine thanks. Things are great/not too bad. You: How's it going? How's life? Reply: Good. What about yourself?	Bei einem zweiten, weniger förmlichen Treffen: Sie: Wie geht es Ihnen? Antwort: Gut, danke. Die Dinge laufen gut/nicht schlecht. Sie: Wie geht's Ihnen? Antwort: Gut. Und Ihnen ?
Introduction	**Vorstellungsrunde**
My name's John Shark. I'm John Shark.	Mein Name ist John Shark.
Mr Worner, let me introduce Ms Eastman. (Nowadays we use Ms to address a woman, which eliminates the need for Mrs or Miss.)	Herr Worner, darf ich Ihnen Frau Eastmann vorstellen.
Mr Worner, I'd like you to meet Ms Eastman.	Herr Worner, ich würde Ihnen gern Frau Eastman vorstellen.

Offering a seat	**Einen Platz anbieten**
Mr Worner, please take a seat.	Bitte nehmen Sie Platz, Herr Worner.
Do sit down, Mr Worner.	Setzen Sie sich doch, Herr Worner.
Mr Worner, would you like to/care to take a seat?	Herr Worner, möchten Sie sich gerne setzen?
Would you mind sitting in that seat?	Würde es Ihnen etwas ausmachen, auf diesem Stuhl Platz zu nehmen?
Making a little small talk	**Smalltalk**
I hope/trust you had a pleasant/comfortable journey over.	Ich hoffe, Sie hatten eine angenehme Anreise?
Did you have a good trip here?	Hatten Sie eine angenehme Anreise hierher?
You came by car/train/plane/taxi, didn't you?	Sie kamen mit dem Pkw/Zug/Taxi, richtig?
Did you have a good flight/crossing?	Hatten Sie einen gute Flug/Überfahrt?
There weren't too many problems at the airport, I hope.	Ich hoffe, es gab nicht zu viele Probleme am Flughafen.
You've obviously been to England before, haven't you?	Sie waren offensichtlich schon mal in England, nicht wahr?
I trust my colleague/assistant/secretary has taken good care of you?	Mein Kollege/Assistent/Sekretärin hat sich hoffentlich gut um Sie gekümmert.
Presenting the company	**Präsentation des Unternehmens**
The company was founded in 1954.	Das Unternehmen wurde 1954 gegründet.
The company with 10 employees back then has become a global group with a workforce of more than 3,500	Aus der kleinen Firma von damals, die mit 10 Angestellten begann, ist heute ein globaler Konzern mit mehr als 3.500 Mitarbeitenden entstanden.
Since 2005, we have been the world market leader in …	Seit 2005 sind wir Weltmarktführer in Bereich der …
The department you have applied for, is …	Die Abteilung, für die Sie sich beworben haben, ist …
General questions	**Allgemeine Fragen**
What's your nationality?	Welche Staatsangehörigkeit haben Sie?
When is your date of birth?	Wann sind Sie geboren?
What's your marital status at the moment?	Wie ist Ihr momentaner Familienstand?
What professional experience do you have in …?	Welche Berufserfahrung haben Sie in …?

Phase 2: Der Bewerber präsentiert sich

Candidate: Describing qualifications	Bewerber: Beschreibung der Qualifikationen
I did a course in Business Administration/economics at university.	Ich belegte einen Kurs in Betriebswirtschaft/Wirtschaftswissenschaften an der Universität.
I did a degree in ... at the university of ...	Ich erlangte den Abschluss in ... an der Universität in ...
I got a degree from the technical college.	Ich erwarb einen Abschluss an der Technischen Hochschule.
I completed a bachelor's degree in ..., a master's degree in ..., a doctorate in ...	Ich schloss mit einem Baccalareus Abschluss in ..., einem Magisterabschluss, einer Doktorarbeit in ... ab.
I obtained GCE A levels in German/Science/Biology/Physics etc.	Ich habe mein Abitur in Deutsch/Biologie/Physik gemacht.
I was awarded a doctorate in ...	Mir wurde der Doktortitel in ... verliehen.
I trained as a commercial clerk.	Ich habe eine kaufmännische Ausbildung gemacht.
At university I specialised in Marketing/Personnel Development/Economics	An der Universität habe ich mich in Marketing/Personalenwicklung/Volkswirtschaft spezialisiert.
Candidate: Describing (English) language skills	**Bewerber: Beschreibung des Sprachlevels**
Well, actually, I owe my knowledge to ...	Ich verdanke mein Wissen dem ...
I learned my English at university/evening classes/in high school etc	Ich habe Englisch in der Universität/Abendschule/Gymnasium gelernt.
I acquired my level of spoken English by staying in England/the USA for a year.	Ich erwarb meine englischen Sprachkenntnisse durch einen einjährigen Aufenthalt in England/USA.
I did a placement in Australia and that helped me to ...	Ich absolvierte ein Praktikum in Australien, was mir geholfen hat zu ...
I studied for the Chamber of Commerce English test.	Ich bereitete mich für den Test der englischen Handelskammer vor.
I studied grammar with a private teacher/tutor.	Ich lernte die Grammatik mit einem Privatlehrer.
I grew up bi-lingual.	Ich wuchs zweisprachig auf.
I'm of English origin.	Ich bin englischer Abstammung.
My written knowledge/skills is/are better than my spoken skills.	Meine schriftlichen Kenntnisse sind besser als meine mündlichen Fähigkeiten.

According to the tests I have an intermediate/upper-intermediate/advanced level of English.	Gemäß den Tests habe ich durchschnittliche/leicht überdurchschnittliche/fortgeschrittene Englischkenntnisse.
I'm proficient in the language.	Ich beherrsche die Sprache.
I'm a relative/false beginner.	Ich bin ein Anfänger mit geringen Vorkenntnissen.
Checking back to see if you have understood	**Rückfragen stellen**
So if I've understood correctly you …	Wenn ich Sie richtig verstanden habe, …
Well, if I've got it right you …	Wenn ich Sie richtig verstanden habe, …
Okay. To put it in a nutshell you …	Ok! Kurz gesagt …
To put it briefly we …	Um es kurz zu sagen, wir …
Let me just summarise that. So you …	Lassen Sie mich das kurz zusammenfassen. Sie …
In other words you're saying I should …	Anders ausgedrückt, Sie sagen, ich sollte …
Just to get it straight, you want me to …	Nur um das klarzustellen, Sie wollen, dass ich …

Phase 3: Fragen zum Lebenslauf und zur Motivation des Bewerbers (und mögliche Antworten)

Giving initial comments on the application/CV	Feedback zum Lebenslauf
Your CV prompts us to question a few things.	Ihr Lebenslauf hat ein paar offene Fragen für uns aufgeworfen.
And, we have to clarify a few points.	Und wir müssen ein paar Punkte klären.
As far as we can see from your resume, you …	Soweit wir Ihrem Lebenslauf entnehmen können …
So, we'll need to ask you to answer a few questions.	Daher möchten wir Sie bitten, ein paar Fragen zu beantworten.
Your application puts us in a quandary.	Ihre Bewerbung hat uns in eine Zwickmühle gebracht.
Well, we had a look at your application, and we think the following: …	Wir haben uns Ihre Bewerbung angeschaut und meinen, dass …
We've checked your CV and we are of the following opinion: …	Wir haben Ihren Lebenslauf geprüft und sind folgender Auffassung …

Why would you like to change your job?	**Warum wollen Sie den Job wechseln?**
It's time for a career change.	Es ist Zeit für ein Karrierewechsel.
I feel I need a challenge	Ich denke, ich brauche eine neue Herausforderung.
I've been in this position three years now and ...	Ich bin jetzt seit drei Jahren auf dieser Position und ...
Why did you leave your last job?	**Warum haben Sie Ihren letzten Arbeitsplatz gekündigt?**
I felt I needed new tasks.	Ich fühlte, dass ich neue Aufgaben brauchte.
I didn't like the atmosphere very much.	Ich habe die Stimmung nicht sehr gemocht.
Well, actually, I was being bullied.	Also, um ehrlich zu sein, ich wurde gemobbt.
What was your best achievement in your last job?	**Was waren die größten Erfolge in Ihrem letzten Arbeitsverhältnis?**
Well, I successfully reorganised the purchasing department.	Nun, ich habe erfolgreich die Einkaufsabteilung reorganisiert.
I successfully introduced some new accounting software.	Ich führte erfolgreich eine neue Buchhaltungssoftware ein.
I managed to get integrated in the team.	Ich konnte mich im Team integrieren.
What didn't you like about your last job?	**Was hat Ihnen an Ihrem letzten Job nicht gefallen?**
I had no contact to overseas staff.	Ich hatte keinen Kontakt zu den im Ausland arbeitenden Kollegen.
There was too much paperwork.	Es gab zu viel Papierkrieg.
Some colleagues were difficult to work with.	Mit einigen Kollegen konnte man schwer zusammenarbeiten.
What is your best qualification for the job?	**Was ist Ihre beste Qualifikation für diese Stelle?**
I think it is my communication skills.	Ich glaube, meine kommunikativen Fähigkeiten.
Probably the fact that I'm very accurate.	Sicherlich die Tatsache, dass ich sehr genau bin.
That I'm really motivated to achieve with this company.	Dass ich wirklich motiviert bin, in diesem Unternehmen was zu erreichen.
What do you find most interesting about this job?	**Was ist für Sie das Interessanteste an dieser Stelle?**
That I'd need to travel a lot.	Dass ich viel unterwegs sein müsste.
That I'll be expected to work independently.	Dass selbstständiges Arbeiten erwartet wird.
That the tasks are clearly defined.	Dass die Aufgaben klar definiert sind.

How do you see yourself in five years?	**Wie sehen Sie sich in 5 Jahren?**
I'd like to be in management.	Ich würde gern im Management sein.
I want to have been promoted to …	Ich möchte zum … befördert worden sein.
I'd prefer to be training other employees.	Ich würde gerne andere Angestellte trainieren.
And what are you doing to achieve your goals?	**Und was machen Sie, um Ihre Ziele zu erreichen?**
I'm already doing a course in …	Ich mache schon eine Ausbildung als …
I'm learning new skills in …	Ich lerne neue Fähigkeiten in …
I've taken examinations in …	Ich habe Prüfungen in … abgelegt.
Why should we hire you?	**Warum sollten wir Sie einstellen?**
You'd get a motivated colleague.	Sie würden einen motivierten Kollegen bekommen.
Because I consider myself to be reliable.	Weil ich mich selbst für zuverlässig halte.
It's a risk, but it's a risk for both of us.	Es ist ein Risiko, aber es ist ein Risiko für uns beide.
Questions to find out if the applicant really wants the job	**Fragen, um herauszufinden, ob der:dieBewerber:in den Job tatsächlich möchte**
Tell me what you know about our company?	Was wissen Sie über unser Unternehmen?
Why do you want this particular job?	Warum wollen Sie diese bestimmte Stelle?
How do you see yourself five years from now?	Wo sehen Sie sich von jetzt an in fünf Jahren?
Why did you leave your last job?	Warum haben Sie Ihren letzten Arbeitsplatz gekündigt?
As project manager, what would you see as your main aims?	Was würden Sie als Projektmanagerin als Ihre Hauptziele ansehen?
Questions to find out if the applicant will fit in the team	**Fragen, um herauszufinden, ob der:die Bewerber:in ins Team passt**
Do you prefer to work on your own or as a member of a team?	Arbeiten Sie lieber allein oder im Team?
Give an example of how you resolved a difficulty with another member of staff in your present company.	Geben Sie uns ein Beispiel, wie Sie eine Schwierigkeit mit einem Kollegen oder einer Kollegin in Ihrem derzeitigen Unternehmen gelöst haben.
What does it take to work in a team?	Was braucht es, um in einem Team zu arbeiten?
How do you see yourself fitting in here?	Wie passen Sie aus Ihrer Sicht hier rein?
How do you feel about taking instructions from someone?	Wie denken Sie darüber, Anweisungen von jemanden entgegenzunehmen?

Phase 4: Fragen zu den Soft Skills des Bewerbers

Organisational skills	**Organisatorische Fähigkeiten**
How do you feel about personal organisation?	Wie ist Ihre Meinung zum Thema »persönliche Ordnung«?
How do you think a filing system should work?	Wie sollte Ihrer Meinung nach ein Ablagesystem funktionieren?
How would you organise a hypothetical project, say on changing purchasing procedures?	Wie würden Sie die Umsetzung eines Projektes organisieren? Versuchen Sie das an dem Beispiel »Neustrukturierung des Einkaufs« darzustellen.
Analytical skills	**Analytische Kompetenzen**
Could you describe the problem situation and point out the key issues?	Können Sie eine schwierige Situation beschreiben und die Kernprobleme identifizieren?
Can you give me examples of problem situations you had to solve?	Bitte nennen Sie mir Beispiele, wie Sie für verfahrene Situationen Lösungen entwickelt haben.
Tell me/us about an inefficient process and what solution you came up with to improve it.	Nennen Sie einen ineffizienten Prozess und wie Ihr Verbesserungsvorschlag dazu aussah.
Making decisions	**Entscheidungskompetenz**
What experience have you had of decision-making?	Welche Erfahrungen haben Sie in der Entscheidungsfindung?
What difficult decisions have you had to make in the past?	Welche schwierigen Entscheidungen haben Sie in der Vergangenheit treffen müssen?
How did you reach the decision? Tell me/us what steps you took in order to reach your decision.	Wie sind Sie zu einer Entscheidung gekommen? Nennen Sie die einzelnen Schritte bis zu Ihrer Entscheidung.
Socialising	**Sozialkompetenz**
Do you prefer to work alone or in a team?	Arbeiten Sie lieber alleine oder im Team?
Have you had any problems getting on with colleagues? Were there any difficult situations?	Hatten Sie irgendwelche Probleme, mit Ihren Kollegen gut zurechtzukommen? Gab es irgendwelche schwierige Situationen?
If there were the following problems with a colleague, how would you handle it?	Wenn die folgenden Probleme mit einem Kollegen auftauchen würden, wie würden Sie damit umgehen?

Communicating	Kommunikationskompetenz
Do you consider yourself a good communicator? Tell me/us why.	Halten Sie sich selbst für einen guten Kommunikator? Begründen Sie, warum das so ist.
Have you written any reports? Could we see an example, please?	Haben Sie irgendwelche Berichte geschrieben? Könnten wir bitte ein Beispiel sehen?
Did you ever have a communication problem with anybody in your last job? What do you think was the reason behind it?	Hatten Sie jemals ein Kommunikationsproblem mit irgendjemand in Ihrem letzten Arbeitsverhältnis? Was, glauben Sie, war der Grund?
Questions to find out if the applicant can do the job	**Fragen zur Qualifikation**
What qualifications do you have for the job?	Welche Qualifikationen haben Sie für die Stelle?
What experience do you feel you can bring to the job?	Was, glauben Sie, können Sie an Erfahrungen für die Aufgabe mitbringen?
Describe the main areas of responsibility in your present job.	Beschreiben Sie die wichtigsten Verantwortungsfelder Ihres gegenwärtigen Arbeitsplatzes.
What do you enjoy most about the work you do?	Was gefällt Ihnen an der Arbeit, die Sie tun, am meisten?
What do you feel you have achieved in your present job?	Was, glauben Sie, haben Sie in Ihrer jetzigen Arbeitsstelle erreicht?
What can you offer this company?	Was können Sie unserem Unternehmen bieten?

Phase 5: Das Interview beenden und offene Fragen klären

Collecting and answering open questions	Offene Fragen sammeln und klären
Are there any questions you might have?	Gibt es von Ihrer Seite noch offene Fragen?
I feel that we already learned a lot about each other. Is there anything that you are not clear about?	Nach meinem Eindruck haben wir jetzt schon sehr viel voneinander erfahren. Ist Ihnen denn jetzt noch etwas ganz unklar?
Did we miss anything in the interview which we should talk about now?	Haben wir im bisherigen Gespräch etwas übersehen, was wir auf jeden Fall jetzt noch besprechen sollten?

Closing the interview	Interview beenden
I would like to thank you very much for this interesting discussion!	Ich danke Ihnen ganz herzlich für das sehr interessante Gespräch!
Before we say goodbye for today, I'd like to show you the department's premises and introduce you to some colleagues.	Bevor wir uns für heute verabschieden, würde ich Ihnen gerne noch die Räumlichkeiten der Abteilung zeigen und Sie auch einigen Kollegen und Kolleginnen vorstellen.
So, now we went a bit over time, but to be honest I'm starving. Would you like to join us for lunch? You would of course be our guest.	So, nun haben wir doch etwas überzogen, und, ehrlich gesagt, mir knurrt der Magen. Wollen Sie uns vielleicht zum Mittagessen begleiten? Selbstverständlich sind Sie herzlich eingeladen.

Übersicht: Häufige Bewerberfragen und wie Sie darauf antworten

What are the typical duties and tasks?	Was sind die typischen Aufgaben und Arbeiten?
Well, you'd have to ...	Nun, Sie müssten ...
So you'd be expected to	Von Ihnen wird erwartet, dass ...
At first, it would be necessary to ...	Zuerst wäre es notwendig, dass ...
And what about the working hours?	Und wie sieht es mit den Arbeitszeiten aus?
The job requires ...	Der Arbeitsplatz verlangt ...
You'd be required to ...	Es wird verlangt, dass man ...
Normal hours are xx, but you might have to ...	Regelmäßige Arbeitszeit sind xx Stunden aber Sie müssten ...
What's the situation with holiday entitlement?	**Wie sieht es mit dem Urlaubsanspruch aus?**
Company rule is ...	Die Firmenregelung ist ...
It's xx weeks per year.	Es sind xx Wochen im Jahr.
It's the statutory entitlement – xx weeks per annum.	Es ist der gesetzliche Anspruch – xx Wochen pro Jahr.
What about pensions?	**Wie sieht es mit der Altersvorsorge aus?**
You can opt to pay into the company pension scheme.	Sie können sich dazu entscheiden, in die betriebliche Altersvorsorge einzuzahlen.
We have no arrangements in this respect. You'll have to find an alternative.	Wir haben diesbezüglich keine Vereinbarungen. Sie werden sich nach einer Alternative umschauen müssen.
We have a retirement fund.	Wir haben eine Pensionskasse.

Can you tell me the notice arrangements?	Wie sieht die Kündigungsregelung aus?
We have xx months' notice to the end of the month.	Wie haben xx Monate Kündigungsfrist zum Monatsende.
You need to give notice in writing.	Sie müssen schriftlich kündigen.
Notice must be given in writing. xx weeks to the month's end.	Kündigungen bedürfen der Schriftform. xx Wochen zum Monatsende.

5.1.3 Beispieldialog: Interview

Background: Interview between Paul Petersen, who has applied for the job of management trainee, Jane Eastmen from Personnel and Alan Sharp from Marketing.

Paul Petersen: How do you do?

Alan Sharp: Please take a seat. I hope you had a comfortable journey. You came by car, didn't you?

Paul Petersen: Yes, I did. The journey was no problem. I had clear instructions from Mrs Jane Eastman: Thank you.

Alan Sharp: Fine, Now, Mr Petersen, I must say your application seems to be a bit of a problem for us. We hadn't expected anybody from outside Britain.

Jane Eastmen: That's right. And that's why we need to ask you a few questions about your university degree. Could you tell us what courses you took at university?

Paul Petersen: Yes, of course. I studied Business Administration for five years. I specialized in marketing, Advertising, Motivation and Project Management.

Alan Sharp: Five years? That's a lengthy course.

Paul Petersen: Yes. We can choose our courses and pick the date of our exams in Germany.

Jane Eastmen: I see. You know English very well. You didn't learn that at school, did you?

Paul Petersen: Well, actually I owe my fluent English to my friend Bill ... *(fade out)*

Alan Sharp: *(fade in)* ... so that was everything on the question of project management.

Paul Petersen: If I've understood everything correctly you would expect me to work in various departments, collect information and probably suggest improvements.

Alan Sharp: That's right. You'd have to monitor business processes, see how we work, and later, perhaps, manage a department of your own if your work is successful in your trial period …

5.2 Bewerber:innen sicher und objektiv bewerten

Checkliste: Kenntnisse des Bewerbers bewerten
Diese Checkliste bzw. dieser Fragenbogen führt Sie sicher durch ein Bewerbungsgespräch in Englisch:

Checklist for a structured interview		
Name of applicant:		
Job applied for:		
Applicant requirements		
What do you imagine this position/job/post involves/includes?		
What is the reason for your application?		
Which special qualifications do you have for it?		
What are your salary expectations?		
State a few facts you know about this company.		
Give several reasons why you would like to work here.		
Present employment situation		
Are you currently in employment?		
If not, how long have you been out of work?		
What was the reason for you becoming unemployed?		
If you are currently fully employed, what is the reason for your application?		
How much notice do you have to give your present employer?		
Job experience		
Present employer		
Address		
Dates of employment		From To

Checklist for a structured interview		
Current/present job title		
What do your duties involve?		
Job changes/promotions at your current company		
Description of duties in the various jobs		
Details of remuneration		Initial salary Final salary
What did you enjoy most about the work?		
What did you enjoy least?		
Reasons for leaving?		
Last employer		
Address		
Dates of employment		From To
Job title		
What do your duties involve?		
Details of remuneration		Initial salary Final salary
What did you like best about the work?		
What did you enjoy least?		
Reasons for leaving?		
Asking for clarification if there are gaps in the CV		
Have you been out of work in the last three years? If yes, what action did you take to find work?		
Education background/qualifications		
What education or qualifications or training do you have that would help your performance in this job?		
Describe your formal education and qualifications.		
Information of a personal nature		
Are you willing to move for your job?		
Are you able to travel?		
How much time per week/month could you spend travelling?		

Checklist for a structured interview							
Interviewer's ranking							
Rating (1 is the highest score.)		1	2	3	4	5
Personal characteristics						
Personal appearance						
Behaviour						
Communication skills						
Cooperation with interviewer						
Job-related characteristics						
Experience for the job						
Job know-how						
Interaction with colleagues						
Effectiveness						

Overall suitability for this position				
Excellent candidate	Well-qualified	Qualified	Weak	Unsatisfactory

Comments or remarks	
Signature	Date

Dos and Don'ts

+	**Dos**
Check all applications following the same criteria.	
Ask job-related questions.	
Make sure that anybody coming into contact with applicants is informed about the legal obligations concerning discrimination.	
Keep a record of the interview including why the candidate was appointed or not.	

You may ask for: • Name, surname • Marital status • Date of birth • Professional experience • Nationality • Certificates • Disability	
-	**Don'ts**
Ask any questions about the candidate's private life.	
Make any jokes, sexist, racist or biased at all.	
Don't ask about: • Pregnancy • Criminal record (except if applicant applies for a position of trust) • Sexual orientation • General state of health • Religion	

5.3 Vokabeln – Vocabulary input

Nachfolgend finden Sie eine Auflistung wichtiger Fachbegriffe rund um das Thema Bewerbungsgespräche.

English	German
accompany	begleiten
achievement *What are your greatest achievements?*	Leistung *Was sind Ihre größten Erfolge?*
acquire	erwerben
appoint	anstellen
avoid	vermeiden
bank clerk	Bankkaufmann/-frau
bi-lingual	zweisprachig
bully *The employee was bullied by his supervisor.*	mobben *Der Mitarbeiter wurde von seinem Vorgesetzten gemobbt.*
business administration	Betriebswirtschaft
career change	Berufs-/Karrierewechsel

English	German
challenge	Herausforderung
commercial clerk	Sachbearbeiter:in, kaufmännischer:e Angestellter:e
company pension *You'll receive a company pension.*	Betriebsrente *Sie werden eine Betriebsrente bekommen.*
criterion, criteria	Kriterium, Kriterien
decision-making	Entscheidungsfindung
determine	bestimmen
disability	Behinderung
discriminatory *This practice is deeply discriminatory.*	diskriminierend *Diese Vorgehensweise ist zutiefst diskriminierend.*
economics	Volkswirtschaft
entitlement	Anspruch
false beginner	Anfänger:in mit geringen Vorkenntnissen
filing system	Ablagesystem
fluent	fließend
GCE A-levels	Abitur
holiday entitlement	Urlaubsanspruch
ice-breaker	Eisbrecher
in charge *I have the money so I'm in charge.*	das Sagen haben *Ich habe das Geld deshalb habe ich das Sagen.*
insist on *I insist on a second chance.*	bestehen auf *Ich bestehe auf einer zweiten Chance.*
insurance clerk	Versicherungssachbearbeiter
job performance	Arbeitsleistung
job-related questions	berufsbezogene Fragen
keep a record	Protokoll führen
likewise	ebenso
loner	Einzelgänger:in
marital status	Familienstand
normal hours	regelmäßige Arbeitszeit

English	German
notice *I handed in my notice last week.*	Kündigung *Ich habe letzte Woche meine Kündigung eingereicht.*
notice arrangement	Kündigungsregelung
notice in writing	schriftliche Kündigung
obligations; legal ~	gesetzliche Verpflichtung
obviously	offensichtlich
one-sided *This comparison is one-sided.*	einseitig *Dieser Vergleich ist einseitig.*
panel interview	Interview mit mehr als einem Fragesteller
performance *Your performance over the last year was not satisfactory.*	Leistung *Ihre Leistung im letzten Jahr war nicht zufriedenstellend.*
pick sth.	etw. auswählen
placement	Praktikum
predict	vorhersagen
proficient *He is proficient in the English language.*	kompetent, etw. beherrschen *Er beherrscht die englische Sprache.*
prospective *My prospective employer sent me an employment contract.*	zukünftig, potenziell *Mein zukünftiger Arbeitgeber schickte mir einen Arbeitsvertrag zu.*
purchasing clerk	Einkaufssachbearbeiter
purchasing department	Einkaufsabteilung
purchasing procedures	Einkaufsabläufe
put it briefly	kurz gesagt
put it in a nutshell	kurz gesagt
quandary	Zwickmühle
questionnaire	Fragebogen
resume	Lebenslauf
retirement fund	Pensionskasse
sales clerk	Verkaufssachbearbeiter

English	German
statutory entitlement	gesetzlicher Anspruch
technical clerk	technischer Angestellter
tidiness	Ordnung
tool	Instrument, Werkzeug
unstructured	unstrukturiert, chaotisch
widely used *This system of recruitment is widely used in Europe.*	weit verbreitet *Diese Personalbeschaffungsmethode ist weit verbreitet in Europa.*
working hours	Arbeitszeit

5.4 Übungen – Exercise time

Task 1: What is the term for the following definitions?

1. Opportunities there might be in the future = p_______________
2. Interview with several interviewers = p_______________
3. Potential future employer = pr_______________
4. A list of things = r_______________
5. Final qualification from a university = d_______________

Task 2: What is the translation of the word in brackets?

1. We've checked your _______________ (Lebenslauf) and we'd like to invite you to an interview.
2. I trained for three years to become a _______________ (Industriekauffrau).
3. Now, at the end of my presentation I'd like to _______________ (zusammenfassen).
4. We need an _______________ (Eisbrecher) at the beginning of a meeting to loosen things up.
5. The candidate needs to adapt to a new working _______________ (Umgebung).

Task 3: Can you remember the correct preposition?

1. We need to collect information _______________ our future partner.
2. It's a risk _______________ both _______________ us.
3. You need to give notice _______________ writing.
4. You can pay _______________ the company pension scheme.

ARBEITSMITTEL AUF MYBOOK.HAUFE.DE

Auf der Internetseite zum Buch finden Sie weitere Übungen zu diesem Modul sowie praktische Vorlagen, um Bewerberinterviews durchzuführen:

- Einladungsschreiben für ein Bewerberinterview
- allgemeines Absageschreiben, das Sie ganz einfach individuell anpassen können
- Absageschreiben an einen Design-Ingenieur, der sich für eine Position in der Forschungs- und Entwicklungsabteilung beworben hatte.

6 Assessment Center (AC)

Zusammenfassung

Um ein professionelles Assessment Center auf Englisch durchführen zu können, bieten wir Ihnen hier zahlreiche Arbeitsmittel an:

- Ein *Gesprächsleitfaden* für die Vorbereitung eines ACs. Sie dient Ihnen zugleich als Checkliste, damit Sie immer alles Wichtige im Blick haben.
- Zwei konkrete *Übungen*, mit denen Sie die Teilnehmer:innen des ACs testen können. Wichtig ist dabei, dass Sie sich zuvor über die Beurteilungskriterien klar werden. Nutzen Sie dazu die Übersicht »Beurteilungskriterien«.
- Ein *Gesprächsleitfaden* bietet Ihnen Unterstützung, damit Sie in der Schlussphase des ACs den Teilnehmern ein konstruktives Feedback geben können.

Spezifische *Vokabeln* finden Sie am Ende des Moduls sowie etliche *Übungen*, um Ihr neu erlerntes Wissen zu überprüfen. Die Lösungen zu den Übungen finden Sie im Anhang dieses Buches.

Die Meilensteine: Professionelle Assessment Center durchführen

Abb. 6: Professionelle Assessment Center durchführen

6.1 Assessment Center vorbereiten

Im Folgenden geben wir Ihnen einen Überblick darüber, wie Sie ein Assessment Center vorbereiten können. Einige dieser Fragen müssen Sie auch dann klären, wenn Sie einen externen Dienstleister mit der Durchführung des ACs beauftragen. Wir empfehlen Ihnen, sich zu jedem dieser Punkte Notizen zu machen, damit Sie Ihre Vorstellungen von dem AC und wie Sie es durchführen wollen mit Kollegen oder eben auch mit einem externen Dienstleister besprechen können.

Assessment centres are usually used for the selection of personnel or to select employees with a high potential. In contrast to the »structured interview« which is mainly used for personnel selection purposes, assessment centres have one decisive advantage: the diagnostic quality is significantly higher as the assessments are based on an actually observed behaviour. The disadvantage of assessment centres is that they are more work-intensive and thus result in higher costs.	Das Instrument Assessment Center wird in der Regel entweder zum Zweck der Personalauswahl oder zur Auswahl Potenzialträgern eingesetzt. Gegenüber dem »strukturierten Interview«, das zumeist zur Personalauswahl eingesetzt wird, hat das Assessment Center einen entscheidenden Vorteil: Die diagnostische Qualität ist deutlich höher, da sich die Bewertung auf tatsächlich beobachtetes Verhalten stützen kann. Der Nachteil des Assessment Centers besteht jedoch im größeren Aufwand und den damit verbundenen höheren Kosten.
The key difference to the interview is that the applicants not only talk but act in simulated situations.	Der zentrale Unterschied zum Bewerberinterview besteht darin, dass die Bewerber:innen nicht nur erzählen, wie Sie z. B. Projekte geleitet haben, sondern dass solch eine Situation simuliert wird.
Below you find tried and tested exercises for assessing one or more candidates. • interviews • cognitive (critical thinking) test • group discussion exercise • case studies • presentations • in-tray exercises	Wir stellen Ihnen hier praxiserprobte Übungen vor, mit denen Sie einen oder mehrere Kandidaten:innen prüfen können. • Interviews • kognitive Tests • Übungen zu Gruppendiskussionen • Fallbeispiele • Präsentationen • Postkorbübungen

6.1.1 Gesprächsleitfaden: Vorbereitung eines Assessment Centers

Defining objectives	**Ziel klären**
Should one or more applicants from a circle of external applicants be chosen for defined positions?	Das Ziel des Assessment Centers ist es, aus einer Gruppe von externen Bewerbern:innen einen oder mehre Personen für eine zuvor definierte Position auszuwählen.
Should colleagues who are potential leaders be found in the existing pool of employees?	Das Ziel des Assessment Centers ist es, aus einer Gruppe von Kollegen:innen einen odere mehrere auszuwählen, die Potenzial für eine Führungsposition haben.
Should employees from the company be identified for special tasks?	Das Ziel des Assessment Centers ist es, aus einer Gruppe von Kollegen:innen einen oder mehrere auszuwählen, die Potenzial für besondere Aufgaben haben.
Defining requirements	**Anforderungen klären**
It would be best, if we met next Tuesday in order to talk about and decide on the requirements for the new position.	Am besten wird es sein, wenn wir uns am Dienstag zusammensetzen, um die Anforderungen an die neue Stelle zu besprechen und festzulegen.
First we will write a job profile to be able to derive the assessment criteria from it.	Wir erstellen zunächst ein Anforderungsprofil, damit wir daraus die Bewertungskriterien ableiten können.
Defining assessment criteria	**Beurteilungskriterien definieren**
Assessment criteria should refer to the requirements to be met by the candidates.	Die Beurteilungskriterien sollten direkt Bezug nehmen auf die Anforderungen, die an die Kandidaten:innen gestellt werden.
Substantiate the assessment criteria using so-called behavioural anchors.	Konkretisieren Sie die Beurteilungskriterien mittels sogenannter Verhaltensanker.
A behavioural anchor for the criterion »motivation« for managers can, for example, be that they are able to successfully motivate staff again despite a failure, and that he or she is able to motivate staff by looking forward and pointing out new opportunities and tackle them in a constructive manner.	Ein Verhaltensanker für das Kriterium Motivation kann zum Beispiel für eine Führungskraft darin liegen, Mitarbeiter:innen trotz eines Misserfolgs erfolgreich neu zu motivieren, indem sie den Blick nach vorn wendet, auf neue Möglichkeiten aufmerksam macht und diese konstruktiv angeht.
Selecting or creating exercises	**Auswahl bzw. Entwickeln von Übungen**
Exercises should be selected or created in line with the core tasks of the post to be filled.	Die Übungen wählen Sie aus bzw. entwickeln Sie entsprechend der Kernaufgaben der zu besetzenden Position.

The candidates for a managerial position simulate an appraisal interview in a role play. If it is about finding an excellent sales person, the role play should simulate a sales talk.	Potenzielle Kandidaten:innen für eine Führungsposition simulieren in einem Rollenspiel ein Mitarbeitergespräch. Wenn es darum geht, einen:e hervorragenden:e Verkäufer:in zu finden, sollte ein Rollenspiel darin bestehen, ein Verkaufsgespräch zu simulieren.
In a group discussion where only a particular topic is given, you can observe the communication skills, persuasive powers and assertiveness of the participants.	In einer Gruppendiskussion, in der Sie nur das Thema vorgeben, können Sie zum Beispiel das Kommunikationsvermögen sowie die Überzeugungs- und die Durchsetzungsfähigkeit der Teilnehmer:innen beobachten.
Presentation exercises are standard procedures in assessment centres. In addition to the communication and rhetorical skills you can also identify their methodological competencies.	Präsentationsübungen gehören zum Standard eines Assessment Centers. Sie können dadurch neben den kommunikativen und rhetorischen auch die methodischen Kompetenzen der Bewerber:innen erkennen.
Selecting applicants	**Bewerber:in auswählen**
Select the applicants you wish to invite to the assessment centre. A sample invitation letter can be found under »Arbeitsmittel« at mybook.haufe.de.	Wählen Sie die Bewerber:innen aus, die Sie zum Assessment Center einladen wollen. Ein Muster-Einladungsschreiben finden Sie bei den Arbeitsmitteln auf mybook.haufe.de.
Selecting assessors	**Beobachter:in auswählen**
An assessment centre needs good, well-trained assessors. For this task you need more than one assessor to be able to put subjective impressions into perspective.	Besonders wichtig für ein Assessment Center ist, dass Sie gute und geschulte Beobachter:innen zur Hand haben. Sie benötigen für diese Aufgabe immer mehr als nur einen:e Beobachter:in, um dadurch möglichst subjektive Eindrücke relativieren zu können.
Setting up a schedule	**Zeitplan erstellen**
Prepare a detailed schedule for the days the assessment centre is taking place.	Erstellen Sie einen detaillierten Zeitplan für die Tage, an denen Sie das Assessment Center durchführen.

6.2 Assessment Center durchführen

6.2.1 Gesprächsleitfaden: Den Teilnehmern den Ablauf des ACs erklären

Nachfolgend finden Sie eine Reihe von Musterformulierungen, wie Sie Teilnehmern den Ablauf des ACs erklären können.

Introducing the system. Keeping it short.	**Aufmerksamkeit erzielen**
So, here we are!	Also dann …
Well, at last we're ready to begin.	Nun, endlich können wir anfangen.
Now it's time to get down to it.	Jetzt ist es Zeit, an die Sache heranzugehen
Okay, let's get started, shall we?	Okay, lassen Sie es uns angehen.
Describing contents/tests	**Beschreibung der Tests**
I'll give you an overview of what you can expect.	Ich werde Ihnen einen Überblick darüber geben, was Sie erwartet.
You're about to hear what's going to happen.	Nachfolgend erfahren Sie, was auf Sie zukommt.
During the assessment you'll be confronted by …	Im Verlauf der Begutachtung werden Sie mit … konfrontiert werden.
Our testing agenda will include …	Unser Testprogramm wird … umfassen.
During that time you'll get …	Währenddessen erhalten Sie, …
In the programme you'll be expected to …	In dem Programm wird von Ihnen erwartet, dass …
There'll be a combination of … in the programme.	In dem Programm wird es eine Kombination von … geben.
The assessment will consist of the following.	Die Bewertung wird Folgendes umfassen.
Reassuring the attendees	**Beruhigende Worte**
There's no need to be nervous or afraid.	Es gibt keinen Grund, nervös oder ängstlich zu sein.
I don't want you to get upset.	Ich möchte nicht, dass Sie sich aufregen.
Don't worry! It's all quite relaxed.	Keine Sorge! Es ist alles ziemlich entspannt.
We don't need to get hot under the collar. It's all really simple.	Es gibt keinen Grund, aufgeregt zu sein. Alles ist recht einfach.
I just want to make you feel at home with what we're doing.	Ich möchte, dass Sie sich sicher fühlen, bei dem, was wir tun.
Describing the activities	**Beschreibung der Aufgabenstellung**
Your tasks will include …	Ihre Aufgabenstellung umfasst …
You'll have to cope with a mixture of different activities.	Sie werden mit einer Mischung von verschiedenen Aktivitäten zurechtkommen müssen.
There'll be a few things to do, including …	Es werden ein paar Dinge zu tun sein, einschließlich …
You'll be assessed according to …	Sie werden aufgrund von … beurteilt werden.

6.2.2 Beispielrede: Introducing attendees to the assessment system

Background: Short introduction speech by Jeff Jefferson welcoming candidates to the assessment center day.

Ladies and Gentlemen,

I'd like to thank you for coming here and to congratulate you at the same time because the assessment centre is no longer the beginning of the interviewing events, it's close to the end.

So here we are! I want to keep it short so I'll describe in general what you can expect. Well, you're here for two days. During that time you will encounter new people to deal with, various activities both individual and group. You'll be monitored all the time, and at the end your assessors – and there will be four of them – will grade you. There's no need to be nervous or afraid. It's in everybody's interest to identify whether you would really fit the job. It's in your interest, too. Imagine the trouble and stress you'd have, if you got into the wrong job. Better to be sure from the beginning.

Now, what do you have to do? Well, you'll have a mixture of activities. First, there'll be a panel interview, then a personality test, and afterwards a group discussion. We'll also have a case study to see how you manage that. All these activities will be considered together. Then we'll have a debriefing.

6.2.3 Übungen für Assessment Center (Panel-Interview/ Gruppendiskussion)

Übung 1: Fragen für ein Panel-Interview

In einem Panel-Interview sind Kandidat:innen gleichzeitig den Fragen mehreren Fragestellern ausgesetzt. Ziel eines Panel-Interviews ist es, zu testen, wie Bewerber:innen mit der stressigen Situation umgehen und Fragen, die in rascher Abfolge von den verschiedenen Panel-Mitgliedern gestellt werden, beantworten. Typische Fragen, mit denen Sie die Eignung von Jobanwärter:innen testen können, finden Sie hier.

Motivation behind the application	Motivation zur Bewerbung
Why have you applied for this job in particular?	Warum haben Sie sich gerade für diese Stelle beworben?
What made you select this position?	Warum haben Sie sich für diese Position entschieden?

Determinants of the choice of study	**Motivation zur Studienfachwahl**
What made you choose your course of study at university/college?	Warum haben sie sich für dieses Studium an der Universität/Hochschule entschieden?
What were the best/worst things in that course?	Was gefiel Ihnen am besten/schlechtesten an dem Studium?
Expectations from the job	**Erwartungen an die Position**
Exactly what are you looking for in this opening?	Was genau erwarten Sie von dieser Stelle?
Tell me what you want from this position.	Sagen Sie mir, was Sie von dieser Position möchten?
Development of potential	**Potenzialentwicklung**
Regarding your personal profile, which of your skills do you need to improve?	Bezüglich Ihres persönlichen Profils, welche Ihrer Fähigkeiten müssen Sie noch verbessern?
Which of your skills do you need to sharpen up, in your opinion?	An welchen Ihrer Fähigkeiten müssen Sie Ihrer Meinung nach noch arbeiten?
Self-organisation	**Selbstorganisation**
When you have to meet a deadline, how do you make sure you do it?	Wenn Sie eine Abgabefrist haben, wie stellen Sie sicher, dass Sie diese auch einhalten?
What do you do to guarantee your deadlines/promises?	Wie stellen Sie sicher, dass Sie Ihre Fristen/Zusagen einhalten?
Career planning	**Karriereplanung**
How does this position fit into your overall career path?	Wie passt diese Stelle in Ihren beruflichen Werdegang?
In which way does this job help your career path?	Auf welche Weise hilft Ihnen diese Stelle bei Ihrer Karriere?
Reasons for change	**Wechselmotivation**
Why do you want a change of company at the moment?	Warum möchten Sie das Unternehmen wechseln?
What's the main reason for you wanting to change your job at this time?	Worin liegt der Hauptgrund für Ihren Wunsch, die Stelle zu wechseln?
Coping with pressure	**Umgang mit Druck**
What really puts you under pressure in your present job? Give an example. How do you manage it?	Was genau setzt Sie unter Druck in ihrem aktuellen Job? Geben Sie ein Beispiel. Wie werden Sie damit fertig?
Is there anything you find difficult to manage in your current job? What is it and how do you deal with it?	Gibt es etwas, was an Ihrem gegenwärtigen Job schwierig ist. Was ist es und wie werden Sie damit fertig?

Übung 2: Gruppendiskussionen

Group discussions are usually employed to check candidates' competencies such as, rhetoric, communication, cooperation and negotiating skills.

Gruppendiskussion werden in der Regel eingesetzt, um Kandidaten:innen auf Kompetenzen wie Rhetorik, Kommunikation, Kooperation und Verhandlungsgeschick hin zu prüfen.

Group discussions can be run in different ways: for example, you can assign certain roles to the individual participants (e.g. for or against) or you ask one participant to start and moderate the group discussion.

Die Form der Gruppendiskussionen können Sie in unterschiedlicher Weise ausgestalten: Sie können z. B. den einzelnen Teilnehmern bestimmte Rollen zuweisen (z. B. dass sie eine Pro- oder Kontra-Position vertreten) oder Sie beauftragen einen der Teilnehmer:innen, die Gruppendiskussion anzuleiten und zu moderieren.

These activities do not require any specialist knowledge – it's about interaction. Communication fluency is important here.
It is not about one individual participant »winning« the discussion or asserting himself. The aim is rather to arrive at a common result or opinion.

Die folgenden Themenvorschläge setzen keine Fachkenntnisse voraus – es geht um die Interaktion. Flüssige Kommunikation ist hier wichtig.
Es geht daher auch nicht darum, dass ein:e einzelner:e Teilnehmer:in die Diskussion »gewinnt« oder sich durchsetzt. Ziel ist es vielmehr, zu einem gemeinsamen Ergebnis bzw. Standpunkt zu kommen.

The observers will quickly identify which one of the participants is taking an active role through analysing, questioning, clarifying, summarising, integrating, pushing on etc.

Die Beobachter:innen werden schnell erkennen, welche Personen durch Analysieren, Nachfragen, Klären, Zusammenfassen, Integrieren, Vorantreiben usw. im Gespräch eine aktive Rolle einnehmen.

Themenvorschläge für Gruppendiskussionen

Think about start-up companies nowadays. What could be done in the first three years of existence to stabilise a young enterprise?

Denken Sie an ein Start-up-Unternehmen heutzutage: Was kann in den ersten drei Jahren des Bestehens getan werden, um das junge Unternehmen zu stabilisieren?

In England the police are normally not armed. Should a police officer carry a weapon? What other methods might there be for combating crime?	In England sind Polizisten in der Regel unbewaffnet. Sollten Polizisten besser Waffen tragen? Welche anderen Methoden gäbe es, um Kriminalität zu bekämpfen?
Some countries have centralised systems like France or the United Kingdom. Others have federal systems such as Germany. What are the pros and cons? What is the way forward?	Einige Staaten sind zentralistisch organisiert wie zum Beispiel Frankreich oder auch Großbritannien. Andere Staaten haben ein föderalistisches System wie zum Beispiel Deutschland. Was sind die Vor- und Nachteile? Was ist der richtige Weg?
The EU is very expensive and is perhaps not very effective as a unit. What has gone wrong? What should the priority be in the future?	Die Europäische Union ist sehr kostspielig und dabei möglicherweise nicht sonderlich effektiv. Was ist falsch gelaufen? Welche Prioritäten sollten in der Zukunft gesetzt werden?

Dos und Don'ts

So sollten sich Kandidaten:innen verhalten bzw. nicht verhalten. Auf diese Dos und Don'ts sollten Sie achten.

Dos	Don'ts
Be yourself at all times.	Don't talk only to hear the sound of your own voice.
Make as many good, but modest contributions as you can.	Do not go quiet for minutes.
Respect your colleagues in the group and keep in mind that different cultures have different behaviours.	Don't try to dominate because you are more fluent!
Listen to what the others are saying.	

6.2.4 Übersicht: Beurteilungskriterien

Die folgende Checkliste enthält Kriterien zur Bewertung von Kandidaten:innen in Assessement Centern. Anschließend finden Sie einen Beispieldialog zu diesem Thema.

Good manners are important standards of evaluation.	Umgangsformen und Körpersprache sind wichtige Bewertungsmaßstäbe.
Presence	**Präsenz**
Does the candidate have a confident stance during his or her presentation?	Tritt der:die Teilnehmer:in in seiner:ihrer Präsentation selbstbewusst auf?
Is he or she looking for eye contact with the other candidates and observers?	Sucht er oder sie den Blickkontakt zu Mitbewerbern:innen und Beobachtern:innen?
Does the candidate outline the arguments with gestures?	Unterstreicht der:die Kandidat:in Argumente mittels Gestik?
Behaviour during breaks	**Verhalten in Pausen**
Does he or she make small talk during the breaks?	Hält er oder sie in den Pausen Smalltalk?
Gestures indicating stress should be avoided by all candidates.	Stressgesten sollten Kandidaten:innen vermeiden.
Nervous finger-tapping or playing with pencils will be considered as negative by the observers.	Nervöses Fingertippen oder das Spielen mit Stiften registrieren Beobachter:innen als Negativpunkte.
Anyone trying to draw attention to themselves and not allowing others to get a word in edgeways will collect penalty points.	Wer sich zu stark in Szene setzt und keinen anderen zu Wort kommen lässt, sammelt Minuspunkte.
No excessive preparation. This can lead to play-acting.	Keine übertriebenen Vorbereitungen. Das kann zur Schauspielerei führen.
Remember the names of the other competitors. You will collect plus points in the discussion later on.	Merken Sie sich die Namen der anwesenden Personen. Damit sammeln Sie Pluspunkte bei der späteren Diskussion.
Form a good relationship with the other candidates. But most important: be yourself!	Bauen Sie ein gutes Verhältnis zu Mitbewerbern:innen auf. Aber das wichtigste ist: Bleiben Sie Sie selbst!

6.2.5 Beispieldialog: Assessment criteria

Background: Discussion between head of personnel recruitment Sam Bright and the Assessment Center consultant Jack Tanner.

Sam Bright: Thanks for coming today. I'll come straight to the point. I need to know how you're doing your assessment. Just some clues to give me a bit of an idea.

Jack Tanner: I know what you mean. I'll focus on the soft skills and little details we're watching for. So, these things apply more or less to all the exercises. Well, we start off and the first thing that we expect is good manners.

Sam Bright: So you follow the general rules of etiquette?

Jack Tanner: Yes. They will have to deal with trained and educated people later. Next is the candidate's posture – the way they stand. Is he/she self-confident but also relaxed? What's the body language like, especially in the presentation? And concerning body language the candidate's eye contact. Does he/she feel comfortable looking into the assessor's eyes?

Sam Bright: What about gesture and hand movement?

Jack Tanner: Yes, hands should »speak«, too. To emphasise points.

Sam Bright: Are you looking for negative impressions, too?

Jack Tanner: Of course. That plays an important role. For example, nervous finger-tapping or playing nervously with a pen or pointer. And it's very important that the candidate tries to integrate with his or her new colleagues. That's where small talk comes in. The candidate should try to make the other group members comfortable and relaxed.

Sam Bright: And tricks?

Jack Tanner: Well, be relaxed, not over-prepared. Use the first names of the others and most of all, be honest and be yourself.

6.3 Feedback geben

The central principle for a constructive feedback is: Do not judge participants but describe their behaviour!

Der zentrale Grundsatz für ein konstruktives Feedback lautet: Nicht Teilnehmer:innen bewerten, sondern Verhalten beschreiben!

So that you can follow this rule it is important that the observers of the assessment centre take detailed notes on the observations they have made.

Um diese Regel befolgen zu können, ist es wichtig, dass sich die Beobachter:innen des Assessment Centers detaillierte Notizen zu ihren Beobachtungen gemacht haben.

Within the scope of the evaluation phase, you and the observers jointly pool the large number of observations made. The aim is to have core statements for each of the criteria.

Im Rahmen einer Auswertungsrunde fassen die Beobachter:innen und Sie gemeinsam die Vielzahl der einzelnen Beobachtungen zusammen. Das Ziel ist, zu jedem Kriterium, das es zu beobachten galt, eine Kernaussage vorliegen zu haben.

In the feedback discussion you should ask the participant to make a self-assessment: How did he or she perceive his or her performance and behaviour during the exercises.	Im Feedbackgespräch bitten Sie zunächst den:die Teilnehmer:in um eine Selbsteinschätzung: Wie hat er die Übungen, seine Leistung und sein Verhalten wahrgenommen.
In a next step you describe your appraisal and explain it based on the detailed notes about the participant's specific behaviour.	Anschließend beschreiben Sie Ihre Einschätzung und erläutern diese anhand der detaillierten Notizen zu seinem konkreten Verhalten.
Discuss the differences and congruencies in the respective perceptions. Justify your appraisal.	Besprechen Sie Differenzen und Übereinstimmungen in der jeweiligen Wahrnehmung. Begründen Sie Ihre Einschätzung.
In particular if the assessment centre is aimed at making a selection amongst colleagues you need to make sure that you do not embarrass any of the participants when giving feedback.	Insbesondere, wenn es sich um ein Assessment Center handelt, das dazu dient, aus Kollegen eine Auswahl zu treffen, müssen Sie beim Feedback darauf achten, keinen der Teilnehmer:innen zu düpieren.
After the assessment centre the candidate may not have been given any decision about whether or not he or she has been selected. Candidates do expect some feedback on how they have performed. It can be that a debriefing session has been integrated in the programme.	Nach dem Assessment Center wurde dem:der Bewerber:in eventuell noch keine Entscheidung mitgeteilt, ob er oder sie ausgewählt worden ist. Der Teilnehmer erwartet irgendeine Resonanz, wie er oder sie sich geschlagen hat. Es ist möglich, dass eine Abschlussbesprechung in dem Programm integriert ist.
The assessor may make the following comments or similar comments.	Der Gutachter macht möglicherweise folgende oder ähnliche Kommentare.

6.3.1 Gesprächsleitfaden: Teilnehmern konstruktiv Feedback geben

Was Sie sagen können, um im Abschlussgespräch zu den Leistungen der einzelnen Teilnehmer:innen konstruktiv Feedback zu geben, erfahren Sie hier.

Positive feedback	**Positives Feedback**
You have performed well in all exercises.	Sie haben in allen Übungen eine gute Leistung gezeigt.
Particularly in the exercise … your performance was excellent.	Insbesondere bei der Übung … war Ihre Leistung hervorragend.

You had a good day today.	Sie hatten einen guten Tag heute.
Actually, you were fine throughout, to mention just one example ...	Sie waren wirklich insgesamt gut, um nur ein Beispiel zu nennen ...
I'm pleased to let you know that your performance was very convincing in the exercise ...	Ich freue mich, Ihnen sagen zu können, dass Ihre Leistung sehr überzeugend bei der Übung ... war.
This was an outstanding performance.	Das war eine außerordentliche Leistung.
Good but some weaknesses	**Gut mit einigen Schwächen**
The individual exercises were quite good, but exercise ... was weak.	Die Einzelübungen waren ziemlich gut, aber die Übung ... war schwach.
You did well in ..., but the ... exercises need to be improved in some areas. For example your sparring partner was much more convincing.	Sie haben sich gut in ... geschlagen, aber die Übungen ... waren in einigen Punkten verbesserungswürdig. Zum Beispiel war Ihr Sparringspartner deutlich überzeugender.
You were weak in group activities. Did you realise that you were hardly involved in the discussion?	Sie waren in den Gruppenübungen schwach. Ist Ihnen aufgefallen, dass Sie sich an dem Gespräch fast nicht beteiligt haben?
Work on your group exercises. You'll do better next time. I arrived at this conclusion because you were able to clearly motivate the group to go on with the two contributions you made to the discussion. I would have liked to see more of this.	Arbeiten Sie an Ihren Gruppenübungen. Sie machen das beim nächsten Mal besser. Ich komme zu dieser Einschätzung, weil Sie mit den zwei Redebeiträgen, die Sie eingebracht haben, die Gruppe deutlich motiviert haben, weiterzumachen. Davon hätte ich sehr gerne noch mehr gesehen.

6.3.2 Beispieldialog: Assessor comments

Background: This is a discussion between the senior assessor Ben Taylor and the candidate Anne Levine.

Ben Taylor: So, now we've looked at the exercises in total. I should tell you that overall you've performed quite well, though there were some areas of weakness.

Anne Levine: Well, I must say I felt completely nervous throughout. I could hardly get my thoughts together.

Ben Taylor: Oh, that's absolutely normal. It's a new situation, new people, how much should you say, what don't you say, etc. In any case, perhaps we could go through the individual items.

Anne Levine: Right. I might do it better next time.

Ben Taylor: To begin with there was the »present yourself« phase at the beginning. It was okay, but sounded a little bit unnatural. And I'd like to give you a few tips for international business. For example, don't introduce yourself and state your age and marital status. It doesn't go down well. It's even taboo in some cultures.

Anne Levine: Oh, I didn't know that. It's quite normal here.

Ben Taylor: Right, but you have to think about the international arena. Next point was the group work exercise. It was obviously a topic you know a lot about.

Anne Levine: Yes, advertising and copywriting was part of my university course.

Ben Taylor: Yes, that was clear. You had more know-how than the others, but you dominated them with your knowledge. That means you didn't collect other opinions, you didn't give potential colleagues a chance.

Anne Levine: Oh, I didn't see it from that point of view.

6.4 Vokabeln – Vocabulary input

Nachfolgend finden Sie eine Auflistung wichtiger Fachbegriffe rund um das Thema Referenzen und Arbeitszeugnisse.

English	German
advantage	Vorteil
assess	bewerten
assessor	Gutachter
battery of activities *We need to counteract the drop in sales with a battery of activities.*	eine Reihe von Aktivitäten *Wir müssen den sinkenden Verkaufszahlen mit einer Reihe von Aktivitäten entgegenwirken.*
behaviour	Verhalten
body language	Körpersprache
career path *She had a successful career path.*	berufliche Laufbahn *Sie hatte eine erfolgreiche berufliche Laufbahn.*
case study	Fallstudie
civil servant	Beamter

English	German
clue *Give me a clue what to do next.*	Anhaltspunkt *Geben Sie mir einen Anhaltspunkt, was ich als nächstes tun soll.*
cognitive	kognitiv
combat crime	Kriminalität bekämpfen
competence	Fähigkeit; Kompetenz
competitor	Konkurrent
consist of	umfassen; beinhalten
convincing *I'm sorry, your arguments are not convincing.*	überzeugend *Es tut mir leid, aber Ihre Argumente sind nicht überzeugend.*
cope with	zurechtkommen mit
copywriting	Texten
course of study	Studium
deadline; guarantee a ~	Einhaltung eines Termins sicherstellen
deadline; meet a ~ *We are in good time. We will meet the deadline.*	eine Abgabefrist einhalten *Wir liegen gut in der Zeit. Wir werden die Abgabefrist einhalten.*
debriefing	Abschlussbesprechung
draw attention to oneself	sich in Szene setzen
encounter	begegnen; antreffen
evaluation standard	Bewertungsmaßstab
eye contact	Augenkontakt
finger-tapping	Fingertippen
fit the job *With his training he fits the job exactly.*	dem Anforderungsprofil entsprechen *Mit seiner Ausbildung entspricht er exakt dem Anforderungsprofil.*
go down well; it doesn't ~ *It doesn't go down well if you are always late.*	das kommt nicht gut an *Es kommt nicht gut an, wenn Sie immer zu spät kommen.*
hardly *I have hardly any experience in dealing with complaints.*	kaum *Ich habe kaum Erfahrung im Umgang mit Beschwerden.*

English	German
in-tray exercise	Postkorbübung
manners; good ~	gute Manieren
modest *Your salary expectations are rather modest.*	bescheiden *Ihre Gehaltserwartungen sind ziemlich bescheiden.*
monitor	überwachen
nowadays	heutzutage
observer	Beobachter
opinion *In my opinion it won't work.*	Meinung *Meiner Meinung nach wird das nicht funktionieren.*
outline	darlegen
outstanding	herausragend
panel interview	Vorstellungsgespräch mit mehreren Interviewern
personnel development	Personalentwicklung
pool of employees	Mitarbeiter-Pool
posture *Your posture shows you are self-confident.*	Körperhaltung *Ihre Körperhaltung zeigt, dass Sie selbstsicher sind.*
pros and cons	Für und Wider
self-confident	selbstbewusst
sharpen up *I have to sharpen up my knowledge of economics.*	verbessern *Ich muss meine Volkswirtschaftskenntnisse verbessern.*
skills	Fähigkeiten
sound *It sounds unrealistic to me.*	sich anhören *Das hört sich unrealistisch für mich an.*
specialist knowledge	Fachwissen
start-up company	neu gegründete Unternehmen
struggle	sich schwer tun
target deployment	zielorientierter Einsatz
testing agenda *The candidate turned pale when she read the testing agenda.*	Testprogramm *Die Kandidatin wurde bleich, als sie das Testprogramm las.*

English	German
throughout	durchweg
topic	Thema
upset; get ~ *The managing director got upset when he saw the figures.*	sich aufregen *Der Geschäftsführer regte sich auf, als er die Zahlen sah.*
varied	verschiedene
weakness	Schwäche

6.5 Übungen – Exercise time

Task 1: Find the verbs from the following nouns.

1. presentation = ________________
2. variation = ________________
3. interest = ________________
4. consideration = ________________
5. reassurance = ________________
6. specification = ________________
7. assessment = ________________
8. appreciation = ________________
9. summary = ________________
10. introduction = ________________

Task 2: Find the adjectives from the following verbs.

1. interest = ________________
2. select = ________________
3. generalise = ________________
4. individualise = ________________
5. personalise = ________________
6. vary = ________________
7. specialise = ________________

Task 3: Put the correct form of the word in brackets into the phrases.

1. I have been a ________________ to ICS for three years. (*consultation*)
2. We ________________ your coming here today for the presentation. (*appreciate*)
3. My presentation will be in three main ________________. (*partly*)
4. This chart ________________ turnover in two countries. (*comparison*)
5. If I understand your question ________________, you want to ... (*correct*)
6. I know it's difficult to accept this ________________, but ... (*decide*)

Task 4: Which terms match the following definitions?

1. a piece of equipment which supports a presentation = _________________
2. people who listen to a presentation or a broadcast = _________________
3. an oral examination in form of a dialogue = _________________
4. a person responsible for checking the suitability of a candidate = _________________
5. the action of doing something better = _________________

ARBEITSMITTEL AUF MYBOOK.HAUFE.DE

Auf der Internetseite zum Buch finden Sie weitere Übungen zu diesem Modul sowie hilfreiche Arbeitsmittel:

- Wir bieten Ihnen eine Vorlage für die Einladung von Bewerbern zum Assessment Center an. Die Vorlage können Sie individuell anpassen, ausdrucken, speichern oder per E-Mail verschicken.
- Schließlich finden Sie einen kurzen Fragebogen zur Beurteilung des Assessment Centers.

7 Einführung neuer Mitarbeiter:innen

Zusammenfassung
Wie Sie neue Arbeitskräfte auf Englisch im Unternehmen einführen und einarbeiten, zeigen wir Ihnen in diesem Modul, denn: je besser die Einarbeitung neuer Mitarbeiter:innen, desto schneller sind sie in der Lage, produktiv zu arbeiten.

- Mit dem *Gesprächsleitfaden* für eine Einführungsveranstaltung haben Sie alle wichtigen Formulierungen parat.
- Zwei *Gesprächsleitfäden* zum Thema »Erster Arbeitstag« und »Mitarbeiter:innen einarbeiten« zeigen Ihnen, wie Sie vorgehen können, und bieten Ihnen zugleich hilfreiche Formulierungen an.
- Darüber hinaus bieten wir Ihnen *Textbausteine* für ein Mitarbeiterhandbuch und einen Verhaltenskodex an. Diese Unterlagen können Sie neuen Mitarbeiter:innen im Rahmen der Einführung übergeben.

Spezifische *Vokabeln* finden Sie am Ende des Moduls sowie etliche *Übungen*, um Ihr neu erlerntes Wissen zu überprüfen. Die Lösungen finden Sie im Anhang dieses Buches.

Die Meilensteine: Neue Mitarbeiter:innen einführen

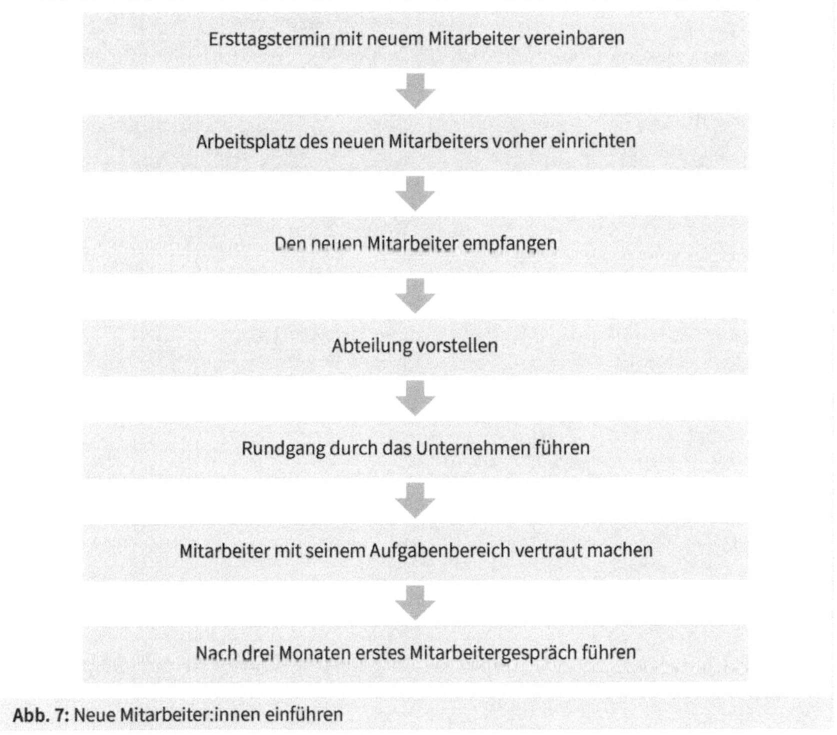

Abb. 7: Neue Mitarbeiter:innen einführen

7.1 Neue Mitarbeiter:innen begrüßen und vorstellen

Make efficient use of induction sessions to present business objectives, operational sequences, products and colleagues of the company to new employees.	Setzen Sie Einführungsveranstaltungen gezielt ein, um neue Mitarbeiter:innen, die Unternehmensziele, die betrieblichen Abläufe, die Produkte sowie selbstverständlich auch die Kollegen rasch näherzubringen.
After presenting the organisation there should be several follow-up discussions to check with the new employees about feedback on the induction and to set goals.	Nachdem die Organisation vorgestellt worden ist, sollten Folgebesprechungen bezüglich der Resonanz der Einführung und zur Zielfestlegung stattfinden.
These meetings can help to identify what training is necessary and to plan future development needs.	Diese Zusammenkünfte können helfen herauszufinden, welche Art von Trainings und zukünftigen Entwicklungsanforderungen notwendig sind.
The induction should be seen as an integral part of a process to understand company objectives and create the job descriptions needed to achieve the goals.	Die Einführung sollte als integraler Bestandteil eines Prozesses zum Verstehen der Unternehmensziele und zur Schaffung der zur Zielerreichung notwendigen Arbeitsplatzbeschreibung gesehen werden.
It should also give information on the physical layout and facilities of the workplace, the key personnel and the company's products and services.	Des Weiteren sollte die Einführung die Weitergabe von Informationen über die Gestaltung und die Einrichtungen des Arbeitsplatzes, Schlüsselpersonal und die Produkte und Dienstleistungen des Unternehmens umfassen.

7.1.1 Gesprächsleitfaden: Einführungsveranstaltung

Musterformulierungen für die Vorstellung und Einführung neuer Arbeitskräfte finden Sie in dem nachfolgenden Leitfaden.

Welcoming the audience	Begrüßung
I'm happy to welcome all of you today.	Ich freue mich, Sie heute alle begrüßen zu dürfen.
It's a pleasure/I'm glad to welcome you here on the occasion of …	Ich freue mich, Sie anlässlich … bei uns begrüßen zu dürfen.
On behalf of the management I'd like to extend a warm welcome to you.	Im Namen der Unternehmensleitung darf ich Sie herzlich begrüßen.

Explaining the reason for the speech	**Anlass der Rede**
And now to get to the reason for our being here.	Und nun zum Grund für unser Zusammentreffen.
So, why are we here today?	Also, warum sind wir heute hier?
Well, what's the purpose of our meeting here?	Nun, was ist der Grund unseres Treffens?
Let's get to the point now.	Lassen Sie uns jetzt zum Kern kommen.
Presenting the new colleague	Neuen Kollegen vorstellen
Ladies and Gentlemen, may I present Mr XY to you?	Meine Damen und Herren, darf ich Ihnen Herrn XY vorstellen?
Dear Colleagues, I'd like to introduce Ms XY to you.	Liebe Mitarbeiter, ich würde Ihnen gerne Frau XY vorstellen.
And here is our new colleague in this department, Ms XY.	Und hier ist unsere neue Kollegin in dieser Abteilung, Frau XY.
I'm sure you've all been waiting to meet Mr XY.	Sicherlich haben Sie alle darauf gewartet, Herrn XY kennen zu lernen.
Giving some information about the new colleague	**Über den neuen Kollegen informieren**
Mr XY has come to us after several years of experience as …	Herr XY ist zu uns gekommen, nachdem er einige Jahre Erfahrung als … gesammelt hat
Ms XY is joining us after a career in France.	Frau XY kommt, nachdem Sie bereits in Frankreich Karriere gemacht hat, zu uns.
Mrs XY has a long history of innovative work in the field of …	Frau XY verfügt über langjährige Erfahrungen im Bereich innovativer Tätigkeiten von …
He has experience in the areas of …	Er hat Erfahrung in den Bereichen …
Her experience includes work in R&D, Sales, etc.	Ihre Erfahrung beinhaltet Tätigkeiten im Verkauf sowie der Forschung und Entwicklung.
Our new colleague has worked as … for several years.	Unser:e neuer:e Kollege:in arbeitet seit mehreren Jahren als …
Describing the new colleague's task	**Aufgaben des neuen Kollegen beschreiben**
Mr XY will be head of department for …	Herr XY wird der Abteilungsleiter von … sein.
Ms XY will be responsible for …	Frau XY wird verantwortlich sein für …
Mrs XY will be in charge of …	Frau XY wird verantwortlich sein für …
Our new colleague is taking over the area of …	Unser:e neuer:e Kollege:in übernimmt den Bereich von …
Her new job will include …	Ihr neuer Arbeitsbereich wird … umfassen.

Closing	Schluss
Well, I'll keep it short ... Best wishes and welcome to	Nun, ich werde es kurz machen. Alles Gute und willkommen bei ...
On behalf of the company, welcome and the best of luck in your new job.	Im Namen des Unternehmens, willkommen und alles Gute für Ihren neuen Arbeitsbereich.
So the best of luck and here's to your success.	Viel Glück und alles Gute für Ihren Erfolg.
I'd just like to finish by saying that we are very happy to have you with us.	Ich möchte damit schließen, Ihnen zu sagen, dass wir sehr froh sind, Sie bei uns zu haben.
That's all for now, I hope you'll feel at home.	Das ist alles für den Augenblick, ich hoffe, Sie werden sich hier wohl fühlen.

7.1.2 Beispielrede: The induction speech

Background: Introducing a new experienced colleague

Ladies and Gentlemen, could I have your attention for a minute, please? Thank you. As you all know I'm not a great person for speeches, I'm more for action, more for getting things done.

And now things have really been »done«. I'm referring to the fact that I am transferring to our headquarters in Bavaria.

And, to come to the reason for our gathering here, I have great pleasure in introducing to you the person who is to fill the gap – if you can call it a gap – after I've gone. He comes to us not, I should mention, from headquarters. He applied to join us directly and in doing so competed with all the applicants on the open market. If I may say so, a laudable act. Laudable, dear colleagues, especially because it is not easy to come to a foreign place to begin work. We all know these are difficult times, what with rationalisation, automation and globalisation. Ladies and Gentlemen, may I present Georg Grosleger?

Mr Grosleger has come to us after several years of successful experience in the United States. He comes highly recommended. His tasks have included supporting the Board of Management with regard to advertising projects. This was in his capacity as assistant to the Board. Now Georg has decided to work on more of his own initiative as head of a department.

He is coming in the »hot« season, to a hot environment to sit on a hot seat. And all of us involved should try to ensure that it does not become an ejector seat. I'm sure our new colleague can count on the support of all of you.

Well, I'll keep it short. On behalf of all of us, I'd like to wish you, Georg, great success for the future in your new role. It is a difficult task, but we believe that you are the right person to do it.

7.2 Vorbereitungen für neue Mitarbeiter:innen

First of all it is necessary to set the tasks and goals for the new employee. The person responsible for the induction should consider what the inductee needs for his work and who his colleagues will be.

Zuallererst ist es notwendig, die Aufgaben und Ziele für die neuen Mitarbeiter:innen festzulegen. Die Person, die für die Einführung verantwortlich ist, sollte prüfen, was die neue Arbeitskraft für seine oder ihre Arbeit benötigt und wer seine oder ihre Kollegen und Kolleginnen sein werden.

His/her first days must be prepared. This includes the internal company rules, e.g. access to e-mails and setting up the workplace …

Bereiten Sie seine:ihre ersten Tage gründlich vor. Dies umfasst Informationen zu den firmeninternen Regelungen, wie z. B. Zugang zum E-Mail-Account sowie die Vorbereitung des Arbeitsplatzes, ….

7.2.1 Gesprächsleitfaden: Der erste Arbeitstag

Hier finden Sie nützliche Redemittel, um neuen Mitarbeiter:innen Kollegen:innen vorzustellen, den Arbeitsplatz zu zeigen, eine Betriebsführung zu machen und den Aufgabenbereich zu erläutern..

Introducing the new employee to his department	**Vorstellung des neuen Mitarbeiters in seiner Abteilung**
If I could have your attention for a moment. I'd like to present … to you.	Wenn ich für einen Moment Ihre Aufmerksamkeit haben könnte. Ich möchte Ihnen … vorstellen.
Okay, Ladies & Gentlemen, I'd like you to meet …	Okay, meine Damen und Herren, ich würde Ihnen gerne … vorstellen.
Can I have everybody's attention? This is Mr XY. He'll be working in our department.	Kann ich bitte Ihre Aufmerksamkeit haben? Das ist Herr XY. Er wird in unserer Abteilung arbeiten.

Could I just introduce you to Mr XY. She's our new colleague.	Kann ich Ihnen Herrn XY vorstellen. Sie ist unsere neue Kollegin.
Fellow purchasers! Perhaps you're wondering who this gentleman is. Well, it's our new	Liebe Kollegen im Einkauf! Vielleicht wundern Sie sich, wer dieser Herrist. Nun er ist unser neue ...
Showing the new employee his workplace	**Neuen Mitarbeiter:innen den Arbeitsplatz zeigen**
So, Ms XY, this is where you'll be working.	So Frau XY, hier werden Sie in Zukunft arbeiten.
Mr XY, let me show you where you'll spend most of your time.	Herr XY, lassen Sie mich Ihnen zeigen, wo Sie die meiste Zeit verbringen werden.
Well, here we are, Mrs XY. I'm sure you know this software.	Nun, Frau XY, hier sind wir. Ich bin mir sicher, Sie kennen diese Software.
I hope you'll feel comfortable with all this.	Ich hoffe, Sie werden sich mit all diesem hier wohl fühlen.
Sorry, we're not state of the art, but we manage just fine.	Entschuldigung, wir sind nicht auf dem letzten Stand, aber wir kommen damit zurecht.
Arranging a company tour for the new employee	**Einen Firmenrundgang für neue Mitarbeiter:innen organisieren**
Ms XY, today we'd be happy to take you round the company.	Frau XY, heute würden wir uns freuen, Ihnen das Unternehmen zu zeigen.
We would like to show you round the company.	Wir möchten Sie im Unternehmen herumführen.
It's time you saw where the real work is done.	Es ist nun an der Zeit, dass Sie sehen, wo die tägliche Arbeit passiert.
It's time for a tour round the company.	Es ist der Tag für eine Führung durch das Unternehmen.
Explaining what the new employee has to do	**Aufgaben neuer Mitarbeiter:innen erläutern**
So, your duties will include ...	Ihre Aufgaben umfassen ...
Well, Ms X, you'll have to take care of the ...	Nun, Frau XY, Sie müssen sich um ... kümmern.
I think you'll be expected to ...	Ich denke, von Ihnen wird erwartet ...
First, you'll need to ..., and then you should ...	Zuerst müssen Sie ... und dann sollten Sie ...

7.2.2 Gesprächsleitfaden: Neue Mitarbeiter:innen einarbeiten

Nachfolgend finden Sie eine Reihe von Musterformulierungen für die Definition neuer Aufgaben von Mitarbeitern:innen, der Überarbeitung von Arbeitsplatzbeschreibungen und Zielvereinbarungen.

Words of welcome at the induction session	Begrüßungsworte zur Einführungsveranstaltung
Ladies and Gentlemen, today I'd like to introduce you to …	Meine Damen und Herren, heute würde ich Ihnen gerne … vorstellen.
Hello! It's my task/job today to present … to you.	Hallo! Es ist heute meine Aufgabe, Ihnen … vorzustellen.
Good morning. Today I want to give you an impression of …	Guten Morgen. Heute möchte ich Ihnen einen Eindruck von … geben.
A warm welcome to you all. Now I'd like to show you why we're all here.	Ein herzliches Willkommen an Sie alle. Nun würde ich Ihnen gerne zeigen, warum wir alle hier sind.
Follow-up meeting to define responsibilities and objectives	**Folgetreffen zur Festlegung der Verantwortlichkeiten und Ziele**
I've asked you to meet me today because …	Ich habe Sie gebeten, sich heute mit mir zu treffen, weil …
We're meeting here again to go through responsibilities and objectives.	Wir treffen uns heute wieder hier, um die Verantwortlichkeiten und Ziele zu besprechen.
Let's discuss what you have to do/manage and what the company requires.	Lassen Sie uns darüber reden, was Ihre Aufgabe ist und was das Unternehmen verlangt.
Let's talk about our company objectives and your responsibilities.	Lassen Sie uns über die Firmenziele und Ihre Verantwortlichkeiten sprechen.
We consider the following to belong in your area of responsibility.	Wir sind der Auffassung, dass Folgendes in Ihrem Verantwortungsbereich liegt.
Agreeing job descriptions	**Vereinbarung der Arbeitsplatzbeschreibung**
Okay, we feel that the following should be included in your job description.	Okay, wir denken, dass das Folgende zu Ihrer Arbeitsplatzbeschreibung gehören sollte.
We understand that times have changed. We'd like to make the following changes …	Wir sind uns darüber bewusst, dass sich die Zeiten geändert haben. Wir würden gerne folgende Änderungen durchführen.
So, if that is okay with you, we can add in …	Nun, wenn das für Sie in Ordnung ist, können wir … hinzufügen.
Well then, we are all in agreement that you will carry out the following jobs: …	Wir sind uns also darüber einig, dass Sie die folgenden Tätigkeiten ausführen werden: …

Documentation	**Dokumentation**
So, we've discussed those points. Could you put them down on paper?	Nun, wir haben über diese Punkte gesprochen. Könnten Sie diese zu Papier bringen?
Okay, could we have that in black and white?	Okay, könnten wir das schwarz auf weiß haben?
Well, we've talked about new tasks, but could you write a one-page report and send it to us for verification?	Wir haben nun über die neuen Aufgaben gesprochen, aber könnten Sie einen 1-seitigen Bericht schreiben und ihn uns zur Überprüfung schicken?
Review of job descriptions	**Überarbeitung von Arbeitsplatzbeschreibungen**
Well, things have changed and we need to look at a few of your tasks ...	Nun die Dinge haben sich geändert und wir müssen uns ein paar Ihrer Aufgaben anschauen.
We've looked at your job description and we need to talk about it.	Wir haben uns Ihre Arbeitsplatzbeschreibung angeschaut und wir müssen darüber reden.
Could we discuss tasks and responsibilities?	Könnten wir über Aufgaben und Verantwortlichkeiten reden?
So, it's time to have look at ...	Nun ist es Zeit, einen Blick auf ... zu werfen.
Revising tasks and goals	**Aufgaben und Ziele überarbeiten**
Mr XX, this meeting is to discuss bringing your job description up-to-date.	Herr XX dieses Treffen dient dazu, Ihre Arbeitsplatzbeschreibung auf den neuesten Stand zu bringen.
In actual fact you're really doing a different job. We need to update your tasks.	In Wirklichkeit machen Sie einen anderen Job. Wir müssen Ihre Aufgaben aktualisieren.
There are a few changes we have to make to your tasks.	Da gibt es ein paar Veränderungen, die wir an Ihren Aufgaben vornehmen müssen.
We're a bit out of date on what you're doing.	Wir sind nicht ganz auf dem Laufenden mit dem, was Sie tun.
Can we talk about it?	Können wir darüber sprechen?
The company has some new goals regarding your job.	Das Unternehmen hat ein paar neue Ziele Ihren Job betreffend.
We'd like to fix/set some new targets/goals.	Wir würden gerne ein paar neue Ziele festlegen.
Let's talk about what we can achieve.	Lassen Sie uns darüber reden, was wir erreichen können.

7.3 Mitarbeiterhandbuch und Verhaltenskodex

The **employee handbook** (or staff handbook) is a document in which the company's practices and procedures are described. This is normally issued to new staff during their induction session. Employees are expected to have read and understood it.

Das **Mitarbeiterhandbuch** ist ein Dokument, in dem die Handlungen und Vorgehensweisen im Unternehmen beschrieben sind. Das Handbuch wird normalerweise an neue Mitarbeiter:innen während der Einführungsphase ausgegeben. Es wird von den Angestellten erwartet, dass Sie es gelesen und verstanden haben.

The company handbook will need to be amended and updated from time to time. It should also be referred to in the contract of employment. Such handbooks may vary according to employer, but in principle the following items should be described and documented:

Das Mitarbeiterhandbuch muss von Zeit zu Zeit erweitert und auf den neuesten Stand gebracht werden. Auf das Mitarbeiterhandbuch sollte auch im Arbeitsvertrag Bezug genommen werden. Solche Handbücher können je nach Arbeitgeber verschieden sein, sie sollten jedoch prinzipiell folgende Punkte beschreiben und dokumentieren:

Introduction	Einführung
Equal opportunities policy	Richtlinie zur Chancengleichheit
health and safety policy	Gesundheits- und Sicherheitsrichtlinie
training	Schulungen
business expenses	Spesen
attendance and timekeeping	Anwesenheit und Zeiterfassung
appearance	äußeres Erscheinungsbild
alcohol and smoking	Alkohol und Rauchen
e-mail and internet	E-Mail und Internet
telephones and other facilities	Telefon und andere Einrichtungen
accepting gifts	Annahme von Geschenken
data protection	Datenschutz
whistle-blowing policy	Richtlinie zur Meldung von Verstößen

dismissal and disciplinary rules and procedure	Kündigungs- und Disziplinarreglungen
grievance procedure	Beschwerdeverfahren
The **code of conduct** describes what is expected of employees. The company will carry out its business in a fair, impartial way, ethically and correctly, with regard to customers, suppliers, employees and the community in general.	Der **Verhaltenskodex** beschreibt, was von den Angestellten erwartet wird. Das Unternehmen wird sein Geschäft in einer objektiven, ethisch einwandfreien und hinsichtlich der Kunden, Zulieferer, Beschäftigten sowie der Allgemeinheit korrekten Art und Weise ausführen.

7.3.1 Textbausteine: Einen Verhaltenscodex verfassen

Nachfolgend finden Sie einige Mustersätze in Form von Dos and Don'ts für die Erstellung eines Verhaltenskodex' für Mitarbeiter:innen.

Dos and Don'ts

+	**Dos**
Do observe all regulations on the use and disclosure of information.	Beachten Sie unbedingt alle Regelungen bezüglich der Verwendung und Bekanntgabe von Informationen.
Observe all rules on protecting information.	Befolgen Sie alle Regelungen hinsichtlich des Schutzes von Informationen.
Make fair dealing a basis for all business activities.	Machen Sie aufrichtiges Handeln zum Grundsatz aller Ihrer geschäftlichen Aktivitäten.
Keep secret all company, client and supplier assets. Use them only for company-approved activities.	Bewahren Sie Stillschweigen über das Vermögen der Firma, Kunden und Zulieferer. Verwenden Sie diese nur für Zwecke, die durch die Firma autorisiert sind.
Apply valid laws, rules and regulations.	Wenden Sie geltendes Recht, Regeln und Bestimmungen an.

Immediately report any illegal or unethical conduct to management.	Melden Sie sofort jede illegale oder unmoralische Verhaltensweise dem Management.
Ask questions, seek guidance, report suspected violations of the code of conduct.	Stellen Sie Fragen, holen Sie Rat ein, melden Sie jeden mutmaßlichen Verstoß gegen den Verhaltenskodex.
- **Don'ts**	
Don't engage in any activity that could create a conflict of interest between you and the company.	Beteiligen Sie sich an keiner Aktivität, die zu einem Interessenskonflikt zwischen dem Unternehmen und Ihnen führen kann.
Don't take advantage of your position to make personal profit.	Nutzen Sie Ihre Position nicht zu Ihrem persönlichen Nutzen aus.
Don't do any insider trading.	Betreiben Sie keinen Insiderhandel.
Don't communicate »closed« information to the public.	Machen Sie keine vertraulichen Informationen öffentlich.
Don't use information obtained in the course of employment for personal gain or benefit.	Verwenden Sie keine Informationen, die Sie im Verlauf Ihrer Beschäftigungen erlangt haben, um sich persönlich zu bereichern oder zu Ihrem persönlichen Vorteil.
Don't take up any outside activities or employment which may have a detrimental effect on the company's interests.	Unterlassen Sie außerdienstliche Aktivitäten oder Beschäftigungsverhältnisse, die sich nachteilig auf die Interessen des Unternehmens auswirken.
Don't discriminate against customers and other employees.	Diskriminieren Sie keine Kunden oder andere Beschäftigte.
Don't receive or give any gift, loan, fee, reward or advantage for doing, or not doing, something or showing favour or disfavour to any person or organisation in your official capacity.	Akzeptieren oder geben Sie kein Geschenk, Darlehen, Gebühren, Belohnung oder Vorteil für Ihr Tun oder Unterlassen oder dafür, dass Sie Sympathien oder Antipathien gegenüber einer Person oder gegenüber einer Organisation im Rahmen Ihrer dienstlichen Eigenschaft zeigen.
Don't use office equipment or mail facilities for personal use.	Benutzen Sie keine Büroausstattung oder Einrichtungen für den Postverkehr für den persönlichen Gebrauch.

7.3.2 Mustertext: Code of Conduct

Code of Conduct

This Code of Conduct outlines behaviour expected from all employees of [Company]. [Company] will conduct its business fairly, impartially, in an ethical and correct manner, and will fully comply with all applicable laws and regulations. In conducting its business, integrity must underlie all Company tasks, including those with customers, suppliers, communities and among employees. The highest standards of ethical business conduct are required of our employees in the performance of their company tasks. Employees may not perform any activity that may bring the Company's honesty, impartiality, reputation into question or otherwise cause embarrassment to the company.

Employees will ensure that:

- they do not engage in any activity that might create a conflict of interest for the company or for themselves as an individual.
- they do not take advantage of their position to seek personal gain through the inappropriate use of non-public information or abuse of their position. This includes not engaging in insider-trading.
- they will follow all regulations on use and disclosure of information. This includes following all guidelines for protecting confidential information.
- they observe that fair dealing is the foundation of all of our business and interactions.
- they will keep secret all company, client and supplier assets and use them only for appropriate company-approved activities.
- without exception, they will comply with all applicable laws, rules and regulations.
- they will immediately report any illegal or unethical conduct to management or other appropriate authorities.
- they will not use information obtained in the course of employment for personal gain or benefit or take up outside activities or employment which may have a detrimental effect on the Company's interests.
- they will not discriminate against customers and other employees
- they will not receive or give any gift, loan, fee, reward or advantage for doing, or not doing, something, or show, in their official capacity, favour or disfavour to any person or organisation.

Every employee has the responsibility to ask questions, seek guidance and report suspected violations of this Code of Conduct. Any form of retaliation against employees who come forward to raise genuine concerns will not be tolerated.

7.4 Vokabeln – Vocabulary input

Nachfolgend finden Sie eine alphabetische Auflistung wichtiger Fachbegriffe rund um die Einführung und Einweisung neuer Mitarbeiter:innen.

English	German
abuse	Missbrauch
access to	Zugriff auf
amend *The contract needs to be amended.*	ändern; überarbeiten *Der Vertrag muss geändert werden.*
appearance	äußeres Erscheinungsbild
attention *She caught the attention of the CEO.*	Aufmerksamkeit *Sie lenkte die Aufmerksamkeit des Vorstands auf sich.*
behalf; on ~ of *I want to thank you on behalf of the management.*	im Namen von *Ich möchte mich bei Ihnen im Namen der Geschäftsleitung bedanken.*
behaviour	Verhalten
best of luck *I wish you the best of luck in your new job.*	viel Glück *Ich möchte Ihnen viel Glück in Ihrer neuen Stelle wünschen.*
briefly	kurz
capacity; in his ~ *I need to sign the contract in my capacity as managing director.*	in seiner Funktion *Ich muss den Vertrag in meiner Funktion als Geschäftsführerin unterzeichnen.*
code of conduct	Verhaltenskodex
company objectives	Unternehmensziele
comply with *We need to comply with the rules.*	befolgen *Wir müssen die Regeln befolgen.*
conflict of interest	Interessenkonflikt
contract of employment	Arbeitsvertrag
data protection	Datenschutz
detrimentally *The exchange rate has affected our results detrimentally.*	schädlich, nachteilig *Der Wechselkurs hat sich nachteilig auf unser Ergebnis ausgewirkt.*
disclosure	Offenlegung
ejector seat	Schleudersitz
embarrassment	Verlegenheit
equal opportunities policy *The firm operates an equal opportunities policy.*	Richtlinie zur Chancengleichheit *Das Unternehmen wendet eine Richtlinie zur Chancengleichheit an.*

English	German
equal opportunity	Chancengleichheit
fee	Gebühr
fill the gap *With his qualifications she certainly fills the gap.*	die Lücke schließen *Mit seinen Qualifikationen schließt sie sicherlich die Lücke.*
follow-up	im Anschluss an etwas
gain *Do not take advantage of this information for your personal gain.*	Bereicherung *Nutzen Sie diese Information nicht zu Ihrer persönlichen Bereicherung.*
gathering	Zusammentreffen
goal	Ziel
grievance procedure *If you have a problem there's a grievance procedure to follow.*	Beschwerdeverfahren *Wenn Sie ein Problem haben, gibt es ein Beschwerdeverfahren, dem zu folgen ist.*
health and safety policy	Gesundheits- und Sicherheitsrichtlinie
impartiality	Unbefangenheit
in principle	prinzipiell
inappropriate	unangemessen
inductee	neuer:e, einzuweisender:e Mitarbeiter:in
induction *The new employee had a one-day induction into the firm.*	Einführung *Der neue Mitarbeiter erhielt eine eintägige Einführung in das Unternehmen.*
innovative	innovativ
integral	integral; wesentlich
issue *We need to issue new shares.*	ausgeben *Wir müssen neue Akten ausgeben.*
job description	Arbeitsplatzbeschreibung
laudable	lobenswert
layout	Ausgestaltung
loan	Darlehen
master something *The team has really mastered the challenge.*	etwas meistern *Das Team hat die Herausforderung in der Tat gemeistert.*
members of staff	Belegschaft; Mitarbeiter

English	German
occasion; on the ~ of *On the occasion of the company's twenty-fifth anniversary you will all receive a bonus.*	anlässlich *Anlässlich des fünfundzwanzigjährigen Jubiläums der Firma erhalten Sie alle einen Bonus.*
on behalf of	im Namen von
one-page report	1-seitiger Bericht
out of date	veraltet
recommend	empfehlen
reputation	Ruf
reward	Belohnung
sacrifice *I sacrificed a lot to be successful.*	opfern *Ich habe eine Menge geopfert, um erfolgreich zu sein.*
staff handbook	Mitarbeiterhandbuch
supplier	Lieferant
take advantage of something	einen Vorteil aus etwas ziehen
update	aktualisieren
valuable	wertvoll
verification	Überprüfung; Bestätigung
violation *This is a violation of company rules.*	Verstoß *Dies ist ein Verstoß gegen die Unternehmensregeln.*
warm welcome *I'd like to express a warm welcome to everyone here.*	herzlich willkommen *Ich möchte alle Anwesenden herzlich willkommen heißen.*
whistle-blowing policy	Richtlinie zur Meldung von Verstößen

7.5 Übungen – Exercise time

Task 1: Find the terms for the following definitions.

1. an introduction to the culture, procedures and processes of a company = i_______________
2. the head office of a company = h_______________
3. the first feelings about something or somebody = f_______________ i

4. a series of rules to ensure a safe workplace = h_______________ & s_______________
5. complaint or dissatisfaction on the part of an employee = g_______________

Task 2: Put in the missing prepositions.

1. It should give information _______________ the physical layout.
2. Step-by-step guide _______________ an induction speech.
3. Mr Smith has come to us _______________ several years of experience in Japan.
4. _______________ behalf of the company, welcome to Mr Smith.
5. I'm sure we can count _______________ the support of all of you.

Task 3: Build up the vocabulary grid.

Noun	Verb	Adjective	Adverb	German
	provide		XXX	
requirement			XXX	
			XXX	erlauben
dismissal			XXX	
	be equal/ equalise			
	be	impartial		

ARBEITSMITTEL AUF MYBOOK.HAUFE.DE

Auf der Internetseite zum Buch finden Sie weitere Übungen zu diesem Modul. Zudem bieten wir Ihnen eine Checkliste und ein Protokoll für die Einarbeitung eines neuen Mitarbeiters sowie ein Muster für ein Mitarbeiterhandbuch.

8 Arbeitsvertrag erstellen

Zusammenfassung

In diesem Modul geben wir Ihnen alle notwendigen Hilfsmittel an die Hand, um einen Arbeitsvertrag auf Englisch zu erstellen:

- Eine *Checkliste* informiert Sie über den Aufbau eines Arbeitsvertrags.
- Und gleich im Anschluss finden Sie über 70 *Textbausteine* zu allen Themen, die in einem Arbeitsvertrag geregelt werden sollten.
- Darüber hinaus bieten wir Ihnen einen vollständigen *Mustervertrag* an, den Sie als Download auch auf mybook.haufe.de finden.

Alle wichtigen themenspezifischen *Vokabeln* finden Sie am Ende des Moduls sowie etliche *Übungen*, um Ihr neu erlerntes Wissen zu überprüfen. Die Lösungen finden Sie im Anhang dieses Buches.

Die Meilensteine: Einen Arbeitsvertrag erstellen

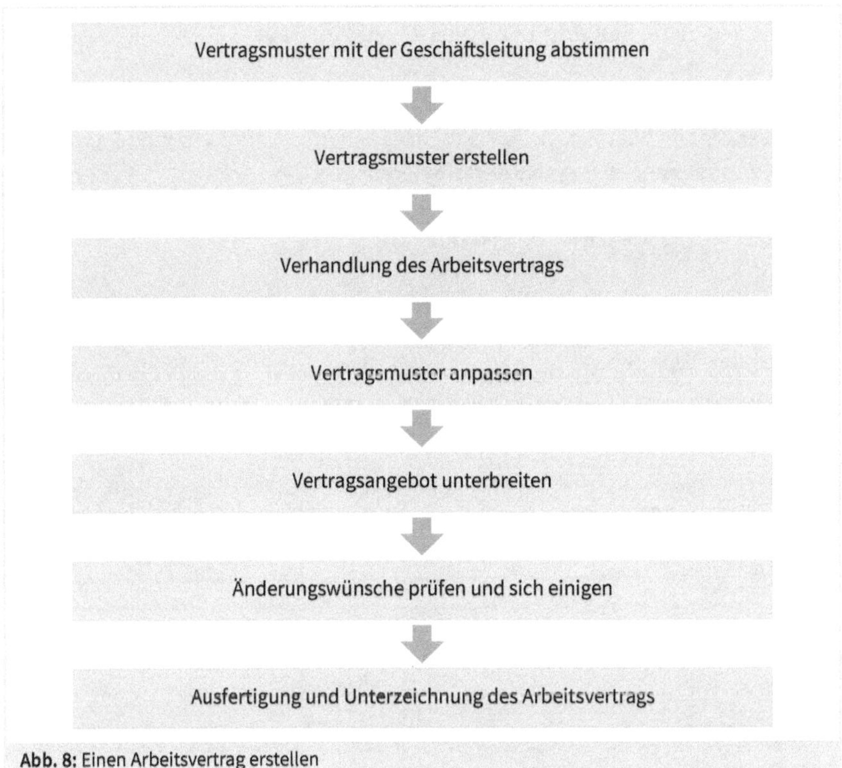

Abb. 8: Einen Arbeitsvertrag erstellen

8.1 Wie ein Arbeitsvertrag aufgebaut ist

The employment contract legally establishes the relationship between the employer and the employee.
The contract can be verbal or in writing or a mix of both of these. Normally the employer has a legal duty to give the prospective employee a written statement of the main terms and conditions under which they are employed. Details vary from country to country.

Der Arbeitsvertrag fixiert in rechtlicher Hinsicht das Verhältnis zwischen dem Arbeitgeber und dem Arbeitnehmer.
Der Vertrag kann mündlich oder schriftlich oder eine Mischung von beidem sein. Normalerweise ist der Arbeitgeber rechtlich dazu verpflichtet, seinen zukünftigen Arbeitskräften eine schriftliche Erklärung über die Bedingungen und Umstände, unter welchen sie beschäftigt werden, auszuhändigen.
Die Einzelheiten sind von Land zu Land verschieden.

8.1.1 Checkliste: Bestandteile eines Arbeitsvertrags

Die nachfolgende Übersicht zeigt die wesentlichen Bestandteile eines englischen Arbeitsvertrags im Überblick. In dem anschließenden Leitfaden erhalten Sie für jeden Abschnitt des Arbeitsvertrags Musterformulierungen.

Checkliste: Bestandteile eines Arbeitsvertrags	
Preamble	Präambel
This agreement is concluded between [Employer] and [Employee].	Dieser Vertrag wird geschlossen zwischen [Arbeitgeber] und [Arbeitnehmer].
This document fixes the terms and conditions of employment under the valid law of [Country].	Dieses Dokument fixiert die Arbeitsbedingungen unter geltendem Recht von [Land].
1. Job Title and commencement	Berufsbezeichnung und Beginn des Arbeitsverhältnisses
2. Remuneration	Vergütung
3. Trial Period	Probezeit
4. Hours of Work	Arbeitszeiten
5. Overtime	Überstunden
6. Holidays	Urlaub

Checkliste: Bestandteile eines Arbeitsvertrags	
7. Sickness	Krankheit
8. Collective agreements	Tarifabkommen
9. 9. Pension	Rente
10. Termination	Beendigung des Arbeitsverhältnisses
11. Confidentiality	Geheimhaltung
12. Non-competition	Wettbewerbsklausel
13. Dismissal, Discipline and Grievance	Kündigung, Disziplinar- und Beschwerdeverfahren
14. Severability clause	Salvatorische Klausel
15. Prior agreements	Vorherige Vereinbarungen
16. Governing Law	Geltendes Recht

8.1.2 Textbausteine: Arbeitsverträge rechtssicher erstellen

Preamble	**Präambel**
This agreement is concluded between [Employer] and [Employee].	Dieser Vertrag wird geschlossen zwischen [Arbeitgeber] und [Arbeitnehmer].
This document fixes the terms and conditions of employment under the valid law of [Country].	Dieses Dokument fixiert die Arbeitsbedingungen unter geltendem Recht von [Land].
1. Job title and commencement	**1. Berufsbezeichnung und Beginn des Arbeitsverhältnisses**
The employer agrees to employ [Name] from [Date] in the capacity of [Job title].	Der Arbeitgeber stimmt der Beschäftigung von (Name) ab (Datum) in der Eigenschaft als (Stellenbezeichnung) zu.
The company will employ [Name] with effect from [Date] as [Job title].	Das Unternehmen beschäftigt (Name) mit Wirkung vom (Datum) als (Stellenbezeichnung).
The employee shall be appointed as [Job title] from [Date].	Der Arbeitnehmer wird mit Wirkung vom (Datum) zum (Stellenbezeichnung) ernannt.
The employee's duties required for this position are set out in the job description.	Die für diese Position erforderlichen Aufgaben des Arbeitnehmers sind in der Arbeitsplatzbeschreibung festgelegt.
The specific duties depend on the individual job description.	Die spezifischen Pflichten ergeben sich aus der individuellen Arbeitsplatzbeschreibung.

The employee will be employed as [Job title] with the following area of responsibility …	Der Arbeitnehmer wird als (Stellenbezeichnung) mit folgenden Verantwortlichkeiten beschäftigt.
The job description may be reasonably modified from time to time to meet the needs of the employer's business.	Die Arbeitsplatzbeschreibung kann von Zeit zu Zeit angemessen modifiziert werden, um den betrieblichen Bedürfnissen des Arbeitgebers gerecht zu werden.
The Company is authorized to change the type and scope of the employee's duties should this become necessary for business reasons.	Das Unternehmen ist ermächtigt, Art und Umfang der Pflichten des Arbeitnehmers bei betrieblicher Veranlassung zu ändern.
The Company reserves the right to extend, reduce or amend the job description, within the scope of the employee's duties.	Das Unternehmen behält sich das Recht vor, die Arbeitsplatzbeschreibung im Rahmen der Aufgaben des Angestellten zu erweitern, zu verkürzen oder zu ändern
2. Remuneration	**2. Vergütung**
The employer shall pay/is obliged to pay a salary of XX per year.	Der Arbeitgeber zahlt ein Jahresgehalt von XX.
The employee shall receive a gross monthly salary of XX.	Der Arbeitnehmer erhält ein Bruttogehalt von XX.
In remuneration of the employee's work, the employee shall receive a gross annual salary of XX.	Als Vergütung für die geleistete Arbeit erhält der Arbeitnehmer ein jährliches Bruttogehalt von XX.
Salary is payable by bank transfer at monthly intervals on the last day of the month.	Das Gehalt wird monatlich mittels Banküberweisung zum letzten Tag eines jeden Monats bezahlt.
The salary is to be paid by bank transfer.	Das Gehalt wird durch Banküberweisung gezahlt.
Remuneration is always payable in the middle of the calendar month.	Die Vergütung wird immer zur Mitte eines Kalendermonats gezahlt.
Should the Company voluntarily award the employee an additional bonus, this shall not give legal entitlement to bonus payments in the future.	Sollte das Unternehmen dem Beschäftigten freiwillige Zulagen gewähren, so entsteht dadurch kein Anspruch auf zukünftige Bonuszahlungen.
The payment of bonuses, shares in profits and other miscellaneous remuneration is made at the discretion of the Company. Regular payment of these does not constitute a legal claim for the future.	Die Zahlung von Zulagen, Gewinnbeteiligungen und anderen Entlohnungen liegt im Ermessen des Unternehmens. Regelmäßige Zahlung dieser begründet keinen zukünftigen Rechtsanspruch.
The right to bonuses is only granted after the employee has been in the Company's service for at least XX months.	Der Anspruch auf Zulagen wird erst nach einer Zugehörigkeit von mindestens XX Monaten gewährt.

3. Trial period and termination of employment during the trial period	**3. Probezeit**
The first XX months are a trial period/probationary period.	Die ersten XX Monate gelten als Probezeit.
This contract is subject to a probationary period of XX months.	Der Vertrag unterliegt der Probezeit von XX Monaten.
The employee is obliged to work XX months as a trial period.	Der:Die Arbeitnehmerin ist verpflichtet XX Monate zur Probe zu arbeiten.
During this period, either contract party may terminate the contract by giving one month's notice to the month end.	Während dieser Zeit kann jede Vertragspartei den Vertrag mit einer Frist von einem Monat zum Monatsende kündigen.
4. Hours of work/hours of employment	**4. Arbeitszeiten**
The employee's normal hours shall be from [Time] to [Time] on Monday to Friday.	Die gewöhnliche Arbeitszeit des Beschäftigten ist montags bis freitags von (Zeit) bis (Zeit).
Regular working hours are XX hours per week. The start and end time of daily hours of work, as well as break entitlements, shall be agreed between the employee and the employee's superiors.	Die Regelarbeitszeit beträgt XX Stunden pro Woche. Der Arbeitsbeginn und das Arbeitsende sowie der Anspruch auf Pausen wird zwischen Mitarbeiter:in und Vorgesetzten vereinbart.
5. Overtime	**5. Überstunden**
The employee may be required to work such additional hours as is reasonable to meet the requirements of the employer's business at an overtime rate of XX per hour.	Vom:Von der Arbeitnehmer:in kann verlangt werden, zusätzliche Arbeitsstunden, die für den Geschäftsbetrieb des Arbeitgebers notwendig sind, zu leisten. Hierfür wird ein Überstundensatz von XX pro Stunde zugrunde gelegt.
The employee is obliged to work overtime by arrangement within the framework of the applicable legal working time regulations.	Der:Die Arbeitnehmer:in ist im Rahmen der gesetzlichen Vorgaben verpflichtet, Überstunden abzuleisten.
Overtime is paid in accordance with the applicable Company guidelines or plant agreements.	Überstunden werden in Übereinstimmung mit den entsprechenden Firmenrichtlinien oder Betriebsvereinbarungen vergütet.
6. Holidays/vacation	**6. Urlaub**
The employee shall be entitled to [Number] days' leave/holiday per calendar year at full pay.	Der:Die Arbeitnehmer:in hat einen Urlaubsanspruch von (Anzahl) Tagen pro Kalenderjahr bei voller Bezahlung.
The employee is entitled to XX working days' vacation per calendar year.	Der:Die Arbeitnehmer:in hat einen Urlaubsanspruch von XX Arbeitstagen pro Kalenderjahr.
The Company will grant XX working days' vacation per year.	Das Unternehmen gewährt 30 Arbeitstage Urlaub pro Jahr.

Any leave shall be agreed with the Company but take into account the preferences of the employee.	Der Urlaub ist mit dem Unternehmen unter Berücksichtigung der Wünsche des:der Arbeitnehmer:in abzustimmen.
Vacation entitlement must be taken during the current year.	Der Urlaubsanspruch muss im laufenden Jahr geltend gemacht werden.
Holidays must be taken at times convenient to the employer.	Der Urlaub muss unter Berücksichtigung der Belange des Arbeitgebers genommen werden.
The employee is obliged to submit applications for leave in writing to the superiors responsible.	Der:Die Arbeitnehmer:in muss Urlaubswünsche schriftlich bei dem:der verantwortlichen Führungskraft einreichen.
No more than XX weeks holiday must be taken at any one time without permission.	Ohne Genehmigung darf der Urlaub XX Wochen am Stück nicht überschreiten.
7. Sickness/Time off work due to sickness	**7. Krankheit**
If the employee is absent from work due to sickness or injury, he or she or someone on his or her behalf should inform the employer of the reason for the absence as soon as possible but no later than [XX].	Sollte der:die Arbeitnehmer:in infolge von Krankheit oder Verletzung dem Arbeitsplatz fernbleiben, so muss er oder sie oder ein:e Beauftragter:e den Arbeitgeber über den Grund des Fernbleibens sobald wie möglich informieren, jedoch nicht später als XX.
In the event of sickness or other inability to work, the employee's direct superior is to be informed on the first day of absence.	Im Falle von Krankheit oder einer anderen Arbeitsunfähigkeit muss der:die direkte Vorgesetzte am ersten Tag der Abwesenheit informiert werden.
The employee is obliged to inform the Company immediately of anything that prevents the employee from performing his duties.	Der:Die Arbeitnehmer:in muss das Unternehmen sofort über alles was ihn:sie daran hindert, seinen:ihren Verpflichtungen nachzukommen, informieren.
The Company reserves the right to ask the employee at any stage of absence to produce/provide a medical certificate/note and/or to undergo a medical examination.	Das Unternehmen behält sich das Recht vor, vom:von der Arbeitnehmer:in während der gesamten Abwesenheit ein ärztliches Attest/Krankenschein und/oder eine medizinische Untersuchung zu verlangen.
After more than three days' sickness a sick/medical note issued by a doctor is to be submitted.	Nach mehr als drei Tagen Krankheit muss ein ärztlich erstelltes Attest eingereicht werden.
In the case of illness, the employee is obliged to present a medical certificate confirming the employee's inability to work.	Im Krankheitsfalle muss der:die Arbeitnehmer:in einen Krankenschein einreichen, der die Arbeitsunfähigkeit bestätigt.
The note should contain details of the expected duration of sickness.	Der Krankenschein sollte Angaben über die voraussichtliche Dauer der Krankheit beinhalten.
The note should indicate the anticipated duration of the absence.	Der Krankenschein sollte die voraussichtliche Dauer der Abwesenheit beinhalten.

The employee shall be paid normal remuneration during sickness absence for a maximum of XX weeks in any period of XX months.	Der:Die Arbeitnehmer:in erhält während der krankheitsbedingten Abwesenheit in einem Zeitraum von XX Monaten für die Dauer von maximal XX Wochen das volle Gehalt.
In the event of sickness the most recent base salary will continue to be paid for the duration of XX months.	Im Falle der Erkrankung wird das zuletzt gezahlte Grundgehalt für die Dauer von XX Monaten weiter gezahlt.
Payment of salary in the case of illness shall continue in accordance with the law governing the continued payment of remuneration.	Die Gehaltsfortzahlung im Krankheitsfall erfolgt gemäß den gesetzlichen Bestimmungen zur Gehaltsfortzahlung.
8. Collective agreements	**8. Tarifabkommen**
There are no collective agreements in force directly relating to the terms of your employment.	Es bestehen keine Tarifvereinbarungen die Ihr Arbeitsverhältnis direkt betreffen.
The terms of the collective agreement dated [Date] made between [XX] and [XX] shall be deemed to be included in this agreement.	Die Bestimmungen der Tarifvereinbarungen vom (Datum), die zwischen XX und XY vereinbart wurden, gelten für diese Vereinbarung.
9. Pension/company pension scheme	**9. Rente**
The employee shall be entitled to join the employer's pension scheme.	Der:Die Arbeitnehmer:in ist berechtigt, der Pensionskasse des Arbeitgebers beizutreten.
The Company undertakes to include the employee in a company pension scheme.	Die Gesellschaft verpflichtet sich, den:die Arbeitnehmer:in in die betriebliche Altersvorsorge aufzunehmen.
Details are set out in the staff handbook.	Einzelheiten sind im Mitarbeiterhandbuch festgelegt.
The company shall offer the employee a salary conversion option in accordance with statutory provisions.	Das Unternehmen bietet dem:der Arbeitnehmer:in eine Gehaltsumwandlungsoption gemäß den gesetzlichen Bestimmungen.
10. Termination	**10. Beendigung des Arbeitsverhältnisses**
The employer may terminate this agreement by giving written notice to the employee as follows: • With not less than one week's notice during the first two years of continuous employment. • With not less than a further one week's notice for each full year of continuous employment after the first two years. • With not less than 12 weeks' notice after 12 years of continuous employment.	Der Arbeitgeber kann diese Vereinbarung wie folgt schriftlich kündigen: • Mit einer Kündigungsfrist von mindestens einer Woche innerhalb der ersten zwei Jahre ununterbrochener Beschäftigung. • Mit einer Kündigungsfrist von mindestens einer weiteren Woche pro Jahr ununterbrochener Beschäftigung nach den ersten zwei Jahren. • Mit einer Kündigungsfrist von mindestens 12 Wochen nach 12 Jahren ununterbrochener Beschäftigung.

Either party may terminate by giving XX weeks' notice to the month end.	Jede Partei kann den Vertrag unter Einhaltung einer Kündigungsfrist von XX Wochen zum Monatsende kündigen.
The employee may terminate this agreement by giving one week's written notice to the employer.	Der:Die Arbeitnehmer:in kann diese Vereinbarung schriftliche mit einer Frist von einer Woche gegenüber dem Arbeitgeber kündigen.
The employer may terminate this agreement without motive or payment in lieu of notice in the case of serious or persistent misconduct such as a major breach of the employer's disciplinary rules.	Der Arbeitgeber kann diese Vereinbarung ohne Angabe von Gründen oder Abgeltung jederzeit im Falle eines schwerwiegenden oder hartnäckigen Fehlverhaltens, wie beispielsweise des schweren Bruchs der Verhaltensrichtlinien für Mitarbeiter, kündigen.
11. Confidentiality/obligation to maintain secrecy	**11. Geheimhaltung**
The employee shall not disclose or allow disclosure of any confidential information.	Der:Die Arbeitnehmer:in darf keine vertraulichen Informationen offenbaren oder deren Offenbarung zulassen.
The employee undertakes to maintain secrecy with respect to …	Der:Die Arbeitnehmer:in bewahrt Stillschweigen hinsichtlich …
The employee is obliged to maintain secrecy regarding business and operational secrets, also after leaving the Company.	Der:Die Arbeitnehmer:die ist zur Geheimhaltung hinsichtlich von Betriebs- und Geschäftsgeheimnissen auch nach seinem Ausscheiden verpflichtet.
Furthermore, the employee undertakes to treat as the property of the Company all documents, printed papers and other business papers of any kind whatsoever entrusted to the employee, as well as outcomes of the employee's professional activities and all records.	Darüber hinaus verpflichtet sich der:die Arbeitnehmer:in, alle Firmendokumente, Drucksachen und andere Dokumente gleich welcher Art, die ihm:ihr anvertraut worden sind, sowie die Ergebnisse der beruflichen Aktivitäten des:der Arbeitnehmers:in einschließlich aller Unterlagen als Eigentum des Unternehmens zu behandeln.
12. Non-competition	**12. Wettbewerbsklausel**
For a period of [Number] months/years after the termination of these agreements, the employee shall not solicit or seek business from any clients of the employer who were clients of the employer at the time during the 12 months immediately preceding the termination of the agreement.	Für eine Zeit von (Anzahl) Monaten/Jahren nach Beendigung dieser Vereinbarung unterlässt es der:die Arbeitnehmer:in, sich bei den in den letzten, unmittelbar der Beendigung der Vereinbarung vorangegangenen 12 Monaten gewonnen Kunden des Arbeitgebers zu bewerben oder geschäftlich aktiv zu werden.
If requested to do so by the Company, the employee will be prepared to sign an agreement restricting competition within the meaning of the regulations in §§ 74 ff. of the German Commercial Code.	Auf Verlangen des Unternehmens erklärt sich der:die Arbeitnehmer:in bereit, eine Wettbewerbsvereinbarung gemäß den Regelungen der §§ 74 ff. HGB zu unterzeichnen.

13. Dismissal, discipline, grievance	**13. Kündigung, Disziplinar- und Beschwerdeverfahren**
The rules are set out in the employer's staff handbook.	Die Regeln sind im Arbeitnehmerleitfaden festgelegt.
Either party may terminate the employment contract by giving [Number] months' notice to the month's end.	Jede Partei kann den Arbeitsvertrag mit einer Frist von (Anzahl) Monaten zum Monatsende kündigen.
The right to terminate without notice for important reasons is not affected.	Das Recht zur außerordentlichen Kündigung bleibt unberührt.
14. Severability clause/changes to the contract	**14. Salvatorische Klausel**
If any provision of this agreement should be held to be invalid it shall to that extent be severed and the remaining provisions shall continue to have full force and effect.	Sollte eine Bestimmung dieser Vereinbarung ungültig sein, so soll diese durch eine Bestimmung ersetzt werden, die dem wahrscheinlichen Willen der Vertragsparteien entspricht. Die restlichen Bestimmungen behalten ihre Gültigkeit.
All changes to this contract must be made in writing. Verbal agreements are not valid. Any changes made in writing to individual sections have no effect on the other agreements.	Alle Änderungen dieses Vertrags bedürfen der Schriftform. Mündliche Abreden sind ungültig. Alle schriftlichen Änderungen einzelner Abschnitte beeinflussen die anderen Vereinbarungen nicht.
Should any individual provisions of the contract become inoperative, the validity of the other sections of the contract remains intact.	Sollten einzelne Bestimmungen dieses Vertrags undurchführbar sein, so bleiben die anderen Klauseln des Vertrags davon unberührt.
Should any provision of the contract be or become ineffective, this does not affect the validity of the remaining provisions.	Sollte eine Klausel dieses Vertrags rechtsunwirksam sein oder rechtsunwirksam werden, betrifft das nicht die Gültigkeit der verbleibenden Klauseln.
15. Prior agreements	**15. Vorherige Vereinbarungen**
This agreement cancels and is a substitute for all previous engagements, agreements and arrangements relating to your employment, all of which shall be deemed to have been terminated by mutual consent.	Diese Vereinbarung beendet und ersetzt alle Ihr Arbeitsverhältnis betreffenden vorhergegangenen Verpflichtungen, Vereinbarungen und Maßnahmen, welche einvernehmlich beendet worden sind.
16. Governing law/place of performance and place of jurisdiction	**16. Geltendes Recht**
The place of performance is the location where the liabilities in dispute are to be performed. The place of performance is [XX].	Erfüllungsort ist der Ort, an welchem die streitgegenständlichen Verpflichtungen zu erfüllen sind. Der Erfüllungsort ist (XX).
This agreement shall be construed in accordance with the laws of [XX] and shall be subject to the exclusive jurisdiction of the [XX] courts.	Diese Vereinbarung soll gemäß den gesetzlichen Bestimmungen von (XX) ausgelegt werden und soll der ausschließlichen Gerichtsbarkeit von (XX) unterworfen sein.

8.2 Vertragsarten

There are several different kinds of contracts of employment. These are related to the type of work, the duration of the employment, and the kind of employment relationship desired by both parties.
The different types of contract regarding duration are: contracts for an indefinite period, and fixed term or limited contracts.
Regarding specific tasks: short-term contracts, part-time contracts and service contracts.

Es gibt verschiedene Arten von Arbeitsverträgen. Dies hängt von der Art der Arbeit, der Dauer der Beschäftigung und der von beiden Parteien gewollten Art des Beschäftigungsverhältnisses ab.
Die verschiedenen Arten von Verträgen bezüglich der Vertragsdauer umfassen unbefristete und befristete Arbeitsverträge.
Hinsichtlich spezieller Aufgaben: Kurzzeitarbeitsverhältnisse, Teilzeitarbeitsverhältnisse und Dienstleistungsverträge.

8.2.1 Muster: Unbefristeter Arbeitsvertrag

Contract of Employment
The following contract of employment is concluded between
[Company name]
and
[Name]
§ 1 Scope of duties

Mr/Mrs/Ms [Name] is employed as [XX] starting on [Date].
The employee has been informed in detail about the scope of his/her duties in line with § 81/1 of the Works Council Constitution Act.
The company reserves the right to assign the employee different yet appropriate and reasonable duties within the company in compliance with his/her collectively agreed rights.

§ 2 Secondary employment

Taking on paid secondary employment, which could compromise the employee's work, is only permitted after receiving prior approval from the company.

§ 3 Probationary period

This contract is concluded with a probationary term of xx months and can be terminated with one month's notice to the end of the month within the probationary period. This does not affect the right to terminate the employment contract without notice.

§ 4 Working times

The provisions of the labour contract for the Hessian Metal Industry determine the working times.
The start and end of the working times, and the time and duration of breaks are governed by the works agreement.

§ 5 Remuneration

The employee receives a monthly gross salary of EUR 0000 (in words: XX euros).
The salary is paid on the last day of each calendar month. The payment is made by funds transfer. The monthly gross salary is calculated on the basis of 00 hours per week as follows:
Salary bracket:
Basic salary: ____________EUR
Voluntary bonus: ________EUR
Gross salary: ___________EUR
The voluntary bonus can be settled with future scale salary increases.

§ 6 Extra bonuses

The employee receives bonus payments in line with the applicable collectively agreed scale or company regulations.
Insofar as these are voluntary bonuses, the payments do not grant any legal rights against the company, even when they are made repeatedly without express reservation of voluntariness.
The right to extra bonuses is only valid after the employee has been working for the company for an uninterrupted term of 6 months in a calendar year.
Employees who have handed in their notice or have been dismissed for breach of their employment contract do not receive Christmas bonus when payment is made. Employees leaving the company before March 31 of the coming year do not receive the voluntary component of the extra bonus, or must pay it back.

§ 7 Attachment of salary

It is not permitted to transfer remuneration. In the event of attachment of salary or assignment of salary in breach of contract, the company is authorized to charge a fee to the value of maximum 3 % of the amount of the attachment or assignment of salary to be processed.

§ 8 Costs and expenses

The reimbursement of costs and expenses complies with the corresponding guidelines for costs.

§ 9 Time off work due to sickness

In the event of sickness or other inability to work, the employee's direct superior is to be informed on the first day of absence.
After more than three days sickness a sick note issued by a doctor is to be submitted with details of the expected duration of sickness.

§ 10 Vacation

The provisions of the collective bargaining agreement determine the vacation entitlement. The employee is currently entitled to 30 working days vacation per full year.
The full vacation eligibility can only be asserted after working for the company for six months. The vacation time is fixed in agreement between the employee and the departmental management in the first quarter of each year within the scope of vacation planning. The nature of the business determines when vacation may be taken, although personal wishes will be granted where possible.
Every employee has the right to vacation pay as stipulated in the collective bargaining agreement. This is calculated as follows:
2.4 % of the gross monthly salary times the number of vacation days in the calendar year. In the event of changes to salary, the last 6 months are used as the basis.

§ 11 Professional secrecy

The employee is obliged to maintain secrecy regarding business and company secrets. This also applies after leaving the company.

§ 12 Ban on competition

If requested to do so by the company, the employee will be prepared to sign an agreement restricting competition within the meaning of the regulations in §§ 74 ff. of the German Commercial Code.

§ 13 Notice period

Once the probationary period has expired, the notice period specified in the collective bargaining agreement is valid.
Should the notice periods that the company is obliged to guarantee the employee be extended due to legal provisions, these extended periods apply to both parties to the contract. The employment relationship ends without notice at the end of the month in which the employee becomes 65 years of age.
Notice must be given in writing.
It is not possible to give notice before the agreed starting date at the company. After notice has been given, the company is entitled to relieve employees of any of their duties at any time, although said employees will continue to receive their salary.

§ 14 Contractual penalty

In the event that the employee intentionally does not perform his/her tasks or terminates his/her work in breach of contract, he/she is obliged to pay the company a contractual penalty to the value of an entire month's income. The total income is based on the average earnings in the last 12 months or, in the event of a shorter term of employment, in line with the average earnings during this term of employment or, insofar as the position has not been taken up, the agreed remuneration. The company is also entitled to assert further damages.

§ 15 Cessions

The employee surrenders his/her claims for damages insofar as he/she is injured by a third party and the company continues to pay his/her salary in the event of sickness.

§ 16 Medical check-up

The company is entitled to demand that the employee attends a free-of-charge medical check-up with the company doctor, either before or after starting work. This is to determine his/her ability to perform his/her job and his/her efficiency.

§ 17 Other provisions

All regulations not mentioned specifically in this contract are based on the provisions of the collective pay agreement for the Hessian Metal Industry, the company regulations and the applicable works agreements.

§ 18 Changes to the contract

All changes to this contract must be made in writing. Verbal agreements are not valid. Any changes made in writing to individual sections have no effect on the other agreements.
Should any individual provisions of this contract become inoperative, the validity of the other sections of the contract remains intact.

§ 19 Expiration dates

All claims resulting from the employment relationship are to be asserted in writing by the parties concluding the contract within 6 months of their due date. Should the other party wish to reject these claims, this rejection must be made within 2 months in writing.

§ 20 Miscellaneous

In line with §§ 22 and 23 of the Privacy Act we hereby inform you that we will save your personal details from the contract.
The employee is obliged to inform the human resources department when he/she moves residence. Should an employee fail to do so, all letters will be sent to the last known address and classed as delivered.
After the fifth month of employment, the employee should submit his/her details to allow transfer of his/her capital accumulation benefits.

§ 21 Place of performance and place of jurisdiction

The place of performance is the location of the party contesting liability. If the employee has not established a place of residence within the country, the place of performance is [Town] for both parties.
The place of jurisdiction for both parties to the contract is the labour court for the place of performance.

[Place, Date] Signature employer

I have read and understood the conditions in this contract and agree to them.

[Place, Date] Signature employee

8.3 Vokabeln – Vocabulary input

Nachfolgend haben wir für Sie alle wichtigen Fachbegriffe rund um das Thema »Arbeitsverträge« aufgelistet.

English	German
amend	ändern
anticipate *We anticipate that the figures will improve next month.*	erwarten *Wir erwarten, dass sich die Zahlen im nächsten Monat verbessern werden.*
appoint	ernennen
attachment of salary	Lohnpfändung
bank transfer	Banküberweisung
capacity; in the ~ of	in der Eigenschaft als
cession	Abtretung
collective bargaining agreement	Tarifvereinbarung
confidentiality	Vertraulichkeit
construe	auslegen
contract; conclude a ~ *We concluded a purchasing contract with the supplier.*	einen Vertrag abschließen *Wir haben einen Einkaufsvertrag mit dem Lieferanten abgeschlossen.*
deemed; shall be ~ as *This paragraph shall be deemed as invalid.*	gelten als *Dieser Paragraph gilt als ungültig.*
discretion; at the ~ of *The invitation is at the discretion of the members.*	im Ermessen von *Die Einladung liegt im Ermessen der Mitglieder.*
dismissal	Kündigung
duty	Aufgabe, Pflicht
employment contract	Arbeitsvertrag
entitlement	Anspruch
framework; within the ~ of *This applies to all rules within the framework of this contract.*	im Rahmen von *Dies gilt für alle Regeln im Rahmen dieses Vertrags.*
governing law	geltendes Recht
grievance	Beschwerden
gross salary	Bruttogehalt

English	German
illness	Krankheit
indefinite period *The contract runs for an indefinite period.*	unbefristet *Der Vertrag läuft für einen unbefristeten Zeitraum.*
invalid	ungültig
medical certificate	medizinisches Attest
miscellaneous	sonstiges
misconduct *You are accused of misconduct at the workplace.*	Fehlverhalten *Sie werden des Fehlverhaltens am Arbeitsplatz beschuldigt.*
non-competition	Wettbewerbsverbot
obligation	Verpflichtung
obliged: to be ~	verpflichtet sein
overtime	Überstunden
penalty *If you don't meet the deadline you have to pay a penalty.*	Konventionalstrafe *Wenn Sie die Frist nicht einhalten, müssen Sie eine Konventionalstrafe zahlen.*
persistent	anhaltend
place of jurisdiction	Gerichtsstand
place of performance	Erfüllungsort
probationary period *The probationary period is three months.*	Probezeit *Die Probezeit beträgt drei Monate.*
prospective employee	potenzieller:e Mitarbeiter:in
relationship	Verhältnis
remuneration	Vergütung
right to *You have the right to appeal.*	Anspruch auf; Recht auf *Sie haben das Recht, Widerspruch einzulegen.*
salary bracket	Gehaltsstufe
scope	Umfang
secondary employment	Nebenbeschäftigung
secrecy; maintain ~ *You must maintain secrecy over confidential information.*	Stillschweigen bewahren *Sie müssen Stillschweigen über vertrauliche Informationen bewahren.*
set out	festlegen
severability clause	Salvatorische Klausel

English	German
sickness	Krankheit
solicit *Now is the time to solicit some customers.*	bewerben *Jetzt ist der richtige Zeitpunkt, einige Kunden zu bewerben.*
statutory provisions	gesetzliche Bestimmungen
subject to *This agreement is subject to confidentiality.*	unterliegen *Diese Vereinbarung unterliegt der Vertraulichkeit.*
submit *Please submit your offer by next Monday.*	vorlegen; einreichen *Bitte reichen Sie Ihr Angebot bis nächsten Montag ein.*
trial period	Probezeit
vacation	Urlaub
valid	gültig
voluntary bonus	freiwillige Zulage
witness *I witness that the information is true.*	bestätigen; bezeugen *Ich bezeuge die Wahrheit der Information.*

8.4 Übungen – Exercise time

Task 1: Translate the words in brackets into English.

1. The contract can be ________________ or in writing. *(mündlich)*
2. There is a ________________ period of six months for this position. *(Probe)*
3. The ________________ required for this job include the following: … *(Aufgaben)*
4. ________________ is always in the middle of the calendar month. *(Bezahlung)*
5. The employee may be expected to work such ________________ as is reasonable. *(Überstunden)*

Task 2: Find words in the texts which have a similar meaning to the following.

1. holidays v________________
2. allow g________________
3. boss s________________
4. medical note s________________
5. total salary r________________
6. end t________________
7. on-going c________________
8. secrecy c________________

Task 3: Which is the correct phrase?

1. Paragraph 1 – line/scope of duties
2. Paragraph 5 – remuneration/renumeration
3. Paragraph 7 – Attachment of salary: the company is authorized to change/charge a fee
4. Paragraph 10 – Vacation: the provision of the tariff/collective bargaining agreement
5. Paragraph 16 – medical check-out/check-up
6. Paragraph 20 – miscellaneous/others

ARBEITSMITTEL AUF MYBOOK.HAUFE.DE

Auf der Internetseite zum Buch finden Sie weitere Übungen zu diesem Modul. Zudem bieten wir Ihnen zwei Mustervorlagen für Arbeitsverträge. Ein Musterarbeitsvertrag richtet sich an leitende Angestellte und das zweite Vertragsmuster an gewerbliche Arbeitskräfte. Beide Verträge können Sie individuell anpassen, ausdrucken, speichern und versenden.

9 Mitarbeiterbeurteilung

Zusammenfassung

Das Ziel einer Mitarbeiterbeurteilung ist immer, die Leistungen des Mitarbeiters zu stabilisieren oder zu verbessern. Lesen Sie in diesem Modul, wie Sie Mitarbeiterbeurteilungen auf Englisch professionell durchführen können. Dazu bieten wir Ihnen:

- *Musterfragen* für einen Selbstbeurteilungsbogen sowie ein fertiges Formular
- Einen *Gesprächsleitfaden* für die Mitarbeiterbeurteilung mit einem Beispieldialog
- *Formulierungsbeispiele* für ein konstruktives Feedback
- Ein *Protokollformular* für das Mitarbeiterbeurteilungsgespräch

Und zum Schluss bieten wir Ihnen einige Übungen und alle wichtigen fachspezifischen Begriffe in alphabetischer Reihenfolge an.

Die Meilensteine: Eine Mitarbeiterbeurteilung erfolgreich durchführen

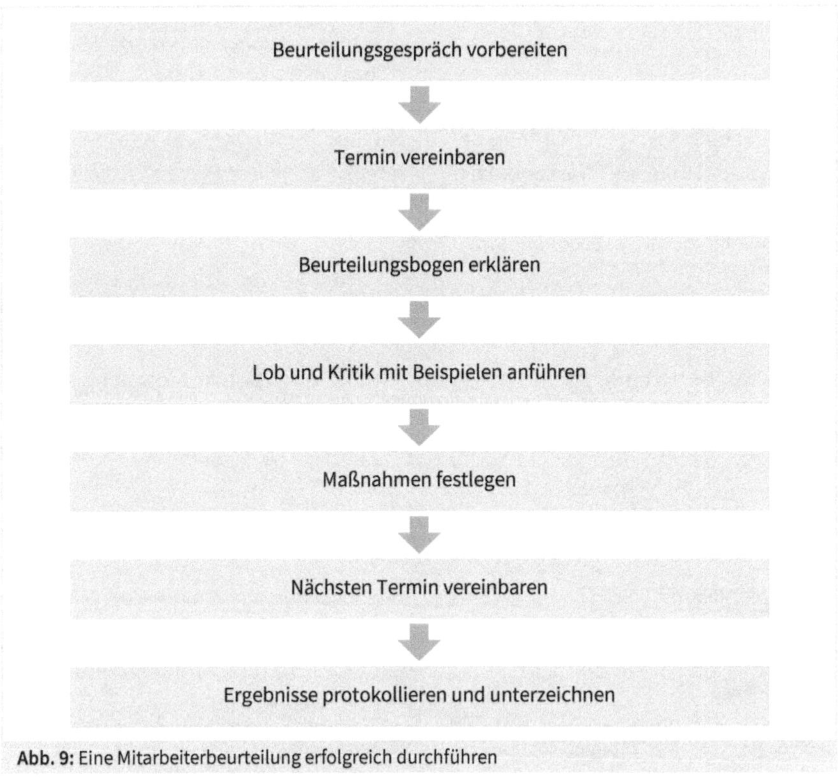

Abb. 9: Eine Mitarbeiterbeurteilung erfolgreich durchführen

9.1 Selbsteinschätzung des Mitarbeiters

The aim of the appraisal is to help employees to recognise, stabilise and further develop their own performance.

There are three types of appraisal. Each one represents a different approach:

- We have a top-down appraisal, led by the appraisee's superior.
- The next is the peer appraisal where employees at the same level assess their colleagues.
- The third type is the 360-degree system. Here, everybody who has been in business contact with the colleague may give his or her opinion.

Often the appraisee is given a self-assessment form in advance of his or her appraisal meeting. The candidate should fill it in while thinking about the low and high points of his or her job.

Das Ziel der Mitarbeiterbeurteilung ist, dass Mitarbeiter:innen die eigenen Leistungen erkennen und diese stabilisieren oder weiterentwickeln können.

Es gibt drei Arten von Beurteilungssystemen, die sich in der Herangehensweise unterscheiden:

- das hierarchische Beurteilungssystem, das durch Vorgesetzte der zu beurteilenden Arbeitskraft geführt wird.
- die gleichrangige Beurteilung, bei der Mitarbeiter:innen der gleichen Hierarchiestufe ihre Kollegen:innen bewerten.
- das 360°-System. Hier kann jeder, der im geschäftlichen Kontakt mit dem:der Kollegen:in stand, seinen oder ihren Kommentar abgeben.

In vielen Fällen erhält die zu berurteilende Person vor dem Mitarbeitergespräch einen Fragebogen zur Selbsteinschätzung.

9.1.1 Textbausteine: Musterfragen für einen Selbstbeurteilungsbogen

In diesem Leitfaden finden Sie eine Reihe von Musterfragen für die Erstellung eines Selbstbeurteilungsbogens.

Key introductory questions	Einleitende Schlüsselfragen
How successfully did you meet your goals for last year?	Wie erfolgreich waren Sie im letzten Jahr bei der Erreichung Ihrer Ziele?
Were you able to achieve your targets for last year?	Waren Sie in der Lage, Ihre für das letzte Jahr gesetzten Ziele zu erreichen?
Did you feel you hit your targets for last year?	Meinen Sie, dass Sie Ihre für das letzte Jahr gesetzten Ziele erreicht haben?
What are you particularly pleased with and what strengths have you shown?	Mit was im Einzelnen waren Sie zufrieden und welche Stärken haben Sie gezeigt?

What satisfied you most and what skills have you displayed?	Was hat Sie am meisten befriedigt und welche Fähigkeiten haben Sie gezeigt?
What made you feel happiest and what were the strong points in your work?	Was hat Sie am meisten gefreut und wo lagen die Stärken bei Ihrer Arbeit?
Talking about improvement	**Über die Weiterentwicklung sprechen**
Where could you develop your performance and how could it be achieved?	Wo konnten Sie Ihr Leistungspotenzial entwickeln und wie konnte das erreicht werden?
In which areas did you succeed in improving your performance and how did you do it?	In welchen Bereichen konnten Sie Ihr Leistungspotenzial verbessern und wie haben Sie das bewältigt?
Were you able to improve the quality of your work and what helped you achieve that?	Konnten Sie Ihre Arbeitsergebnisse verbessern und was hat Ihnen dabei geholfen?
Identifying change in the job, and in the future	**Änderungen der Arbeitsaufgaben und zukünftige Änderungen erkennen**
What changes were there in your role in the last year?	Welche Veränderungen Ihrer Funktion haben letztes Jahr stattgefunden?
How have activities changed in the past XX months?	Wie haben sich die Aktivitäten in den letzten XX Monaten geändert?
What changed about your job in the last year?	Wie hat sich Ihre Arbeit im letzten Jahr geändert?
What do you foresee for the future?	Was, denken Sie, wird die Zukunft bringen?
How does the future look to you regarding …?	Wie sehen Sie bezüglich … die Zukunft?
What do you imagine will change in the future?	Was, denken Sie, wird sich in Zukunft ändern?
Identifying problems at work	**Probleme am Arbeitsplatz erkennen**
What problems did you confront in performing your role last year?	Welche Probleme sind Ihnen letztes Jahr bei Ausübung Ihrer Funktion begegnet?
Which problems did you meet when you carried out your role?	Welche Probleme sind Ihnen bei Ausübung Ihrer Rolle begegnet?
Which issues did you have to deal with in order to do your job?	Mit welchen Themen mussten Sie sich bei Ausübung Ihrer Aufgaben beschäftigen?
Asking about dealing with problems	**Fragen zur Bewerkstelligung von Problemen**
How did you overcome them?	Wie haben Sie diese gemeistert?
In what way did you cope with them?	In welcher Weise haben Sie diese bewältigt?
Which methods did you use to deal with them?	Mit welchen Methoden haben Sie diese bewerkstelligt?

9.1.2 Formular: Fragebogen zur Selbsteinschätzung

Self Assessment Form

Name: ____________________ Job title: ______________________

Date of formal appraisal: __________________

Date of last formal appraisal: _______________

Name of line manager: ____________________

Review of your general progress

____

____

____

Review of your performance

____

____

____

Review of your role

____

____

____

Review of your personal development

____

____

____

9.2 Ein Mitarbeitergespräch effizient durchführen

A typical appraisal interview may have the following structure: building up a positive atmosphere, agreeing the agenda, pushing the communication forward, creating an action plan with the appraisee and finally summarising the agreed action for commitment.

Ein typisches Beurteilungsgespräch kann die folgende Struktur haben: Aufbau einer positiven Atmosphäre, Einigung über die Besprechungspunkte, Vorantreiben der Kommunikation, Erstellung eines Maßnahmenplans mit dem zu Beurteilenden sowie abschließend eine Zusammenfassung der vereinbarten Verpflichtungen.

9.2.1 Gesprächsleitfaden: Mitarbeiterbeurteilung

Dieser Leitfaden zeigt Ihnen den beispielhaften Aufbau eines Mitarbeitergesprächs sowie einige Musterformulierungen für die verschiedenen Abschnitte des Beurteilungsgesprächs.

Establishing a positive purpose to the appraisal	**Eine positive Zielsetzung für das Beurteilungsgespräch schaffen**
I think we had some good ideas at the last meeting, didn't we?	Ich denke, wir hatten im letzten Treffen ein paar gute Ideen.
It was a good year. Let's try to think about what we could improve.	Es war ein gutes Jahr. Lassen Sie uns darüber nachdenken, was wir verbessern können.
Well, I've been looking forward to reviewing recent successes with you.	Ich habe mich darauf gefreut, die jüngsten Erfolge zu besprechen.
I'm sure we can focus on the positive things and set our goals for the next XX months.	Ich bin sicher, dass wir die positiven Dinge herausgreifen und unsere Ziele für die nächsten XX Monate festlegen können.
Agreeing the items for the agenda	**Besprechungspunkte vereinbaren**
So, if it's okay with you I'd like to include …	Wenn es für Sie in Ordnung ist, würde ich gerne … einbeziehen.
Item 4 isn't so vital. Let's agree to delete it.	Punkt 4 ist nicht so entscheidend. Einigen wir uns darauf, diesen zu streichen.
Perhaps we could talk about … and … today.	Vielleicht können wir heute über … und … sprechen.
What are your feelings about today's content?	Wie denken Sie über die heutigen Inhalte?

Getting the communication with the appraisee going	Das Gespräch mit dem zu Beurteilenden in Gang bringen
So, Mr XY, please feel free to suggest items.	Nun, Herr XY, Sie können gerne Themen vorschlagen.
Ms XY, perhaps you would like to sum up your achievements of the past XX months?	Frau XY, vielleicht würden Sie gerne das Erreichte der letzten XX Monate zusammenfasse?
Would you like to comment on this?	Würden Sie gerne dazu einen Kommentar abgeben?
We should do this appraisal in a relaxed way, nothing too formal.	Wir sollten diese Beurteilung entspannt und nicht zu formal durchführen.
Perhaps we should brainstorm a little together.	Vielleicht sollten wir beide zusammen Ideen sammeln.
Putting together an action plan	**Aufstellung eines Maßnahmenplans**
Let's decide on the action to be taken.	Lassen Sie uns darüber entscheiden, was zu tun ist.
Well, from what you have said, I'd suggest the following To Dos.	Nun, von dem, was Sie gesagt haben, würde ich die folgenden abzuarbeitenden Schritte vorschlagen.
Perhaps we could prioritise our action points.	Vielleicht könnten wir unsere Maßnahmen priorisieren.
We need to do … and also set a timeline.	Wir müssen … tun und auch einen Zeitrahmen setzen.
Summarising agreed actions	**Zusammenfassung der vereinbarten Maßnahmen**
Yes, I believe we agreed on the following.	Ja, ich glaube, wir sind uns über das Folgende einig geworden.
We said there were three points to clarify.	Wir haben gesagt, dass drei Punkte zu klären sind.
We agreed there were three decisions to implement.	Wir waren uns darüber einig, dass drei Entscheidungen zu implementieren waren.
So, we both accepted that I need to …	Wir waren uns beide darüber einig, dass ich … muss.
Tracking measures and goals	**Maßnahmen und Ziele nachverfolgen**
Later in the year we should check against the plan.	Im Laufe des Jahres sollten wir den Plan überprüfen.
We need to keep an eye on the goals/objectives.	Wir müssen die Ziele im Auge behalten.

We'll have to revise/adjust our goals in XX months.	Wir werden die Ziele in den nächsten XX Monaten überprüfen/anpassen müssen.
Trust is good but control is better.	Vertrauen ist gut, Kontrolle ist besser.
Offering support	**Unterstützung anbieten**
Is there anything I can do to support you in your tasks?	Kann ich etwas für Sie tun, um Sie bei Ihren Aufgaben zu unterstützen?
Can I do anything to help you perform better?	Kann ich Ihnen irgendwie helfen, um Ihre Leistungen zu verbessern?
Just let me know what I can do to make it easier for you.	Lassen Sie mich wissen, was ich tun kann, um es leichter für Sie zu machen.
Giving the employee the opportunity to criticise personal interaction	**Mitarbeiter:innen die Möglichkeit geben, den Umgang miteinander zu kritisieren**
Am I doing anything that hinders your performance?	Gibt es etwas, was ich tue, dass Sie daran hindert, Ihre Leistung zu erbringen?
Am I holding you back in any way?	Hemme ich Sie in irgendeiner Weise?
Am I being unclear in instructions?	Sind meine Anweisungen nicht klar?
Please let me know if you don't feel comfortable with anything I've said.	Bitte, lassen Sie es mich wissen, wenn ich etwas gesagt habe, mit dem Sie nicht zufrieden sind.
Checking for strength and weaknesses	**Stärken und Schwächen aufdecken**
What do you see as your key strengths and weaknesses?	Wo sehen Sie Ihre größten Stärken und Schwächen?
What do you think your plus points are?	Was denken Sie, sind Ihre Pluspunkte?
Where do you feel you might have to improve something?	Wo, glauben Sie, müssen Sie sich noch verbessern?
What are the advantages and disadvantages in your approach to the work?	Was sind die Vor- und Nachteile bei Ihrer Einstellung zur Arbeit?
What can really help you to achieve your objectives?	Was kann Ihnen bei der Erreichung Ihrer Ziele wirklich helfen?
Questions concerning the employee's team qualities	**Fragen zur Teamfähigkeit des Mitarbeiters**
What about improving yourself as a team member?	Was halten Sie davon sich als Teammitglied zu verbessern?
How could you become a better team member?	Wie können Sie zu einem besseren Teammitglied werden?
How can we help you to develop yourself as a team member?	Wie können wir Ihnen helfen, um sich als Teammitglied weiterzuentwickeln?

Checking for areas of improvement	Verbesserungspotenziale aufdecken
How could the job or system be improved?	Wie könnte die Stelle oder das System verbessert werden?
What could we do to refine the system?	Was können wir tun, um das System weiterzuentwickeln?
How can we make the job suit you better?	Wie können wir die Stelle geeigneter für Sie machen?
Where could we fine-tune the system to ...?	Wo können wir das System abstimmen, um ...?

9.2.2 Beispieldialog: Appraisal interview

Background: This is an appraisal interview between the Head of Sales, Wesley Turner, and a sales clerk, Jim Kelly.

Wesley Turner: So Jim, you realise it's exactly a year since our last appraisal interview.

Jim Kelly: It's true. How time flies! And in fact I feel as if I hardly had time to achieve last year's priority targets, and now here we are again.

Wesley Turner: Well, I've been looking forward to reviewing recent goals with you. I mean you pretty well achieved everything and on time, too. Let's single out the positives and set our goals for, say, the next nine months.

Jim Kelly: Okay. Shall we set up things together, or how do you want to do it?

Wesley Turner: If it's okay with you I'd like to include some items on the agenda. For example, number of customer visits per month for the coming year. And, of course, turnover targets for you personally, and for the team that you're a member of.

Jim Kelly: Okay, were we too far short in the last twelve months?

Wesley Turner: No, not at all. Results were fine. You did a fine job. It's new rules from top management and from our shareholders. By the way, we will have to talk about bonus, too.

Jim Kelly: I was expecting that.

Wesley Turner: Well, please feel free to suggest anything you want to talk about. This is a two-way show here.

Jim Kelly: Well, if I might just mention it I'd like to include an item on vacation, holiday allocation and timing.

Wesley Turner: Sure. No problem! We should keep in mind, of course, that the main priority today is for us to set your personal goals in a way that the company and you are both committed and rowing in the same direction.

Jim Kelly: I'll keep that in mind.

Wesley Turner: Once we've agreed targets, personal and corporate goals we should do some brainstorming which should result in an action plan for the two of us, and you realise that I'll be monitoring your success and so on in the next coming months. If we get off track we can come together and adapt your planning. And with regard to that point I'd like to show you the company's latest little tool. It's a Performance Agreement. It's a sort of »commitment« on the part of the employee.

9.3 Konstruktiv Feedback geben

Sometimes the assessor needs to give negative feedback. This must be done carefully and describe exactly and in detail what the assessor was not satisfied with. It is important that the employee has the opportunity to demonstrate his point of view.
Involve the appraisee in finding solutions to problems and weaknesses and planning relevant further educational measures.

Manchmal muss die bewertende Person eine negative Rückmeldung geben. Dabei muss man sehr vorsichtig agieren und ganz konkret und detailliert beschreiben, womit man nicht zufrieden war. Und es ist wichtig, dass Mitarbeiter:innen die Möglichkeit gegeben wird, ihre Sicht darzustellen.
Beziehen Sie die beurteilte Person mit ein, wenn es darum geht, Lösungswege für Probleme und Schwächen zu entwickeln und konkrete Weiterbildungsmaßnahmen zu planen.

9.3.1 Formulierungsbeispiele: Lob und Kritik

Praising	Loben
So, you've had a pretty good year.	Sie hatten ein ziemlich gutes Jahr.
Well, most of your targets were reached, weren't they?	Die meisten Ihrer Ziele wurden erreicht, nicht wahr?

It wasn't all easy but on the whole you've done really well.	Es ist nicht einfach gewesen, aber im Großen und Ganzen haben Sie sich gut geschlagen.
Regarding your targets, it's been a good year.	Hinsichtlich Ihrer Ziele ist es ein gutes Jahr gewesen.
After such a good start we should look at further milestones.	Nach einem solchen guten Start sollten wir auf die nächsten Meilensteine schauen.
Criticising	**Kritisieren**
The last six months were quite good but there are some areas we have to talk about.	Die letzten sechs Monate waren ziemlich gut, aber da gibt es ein paar Bereiche, über die wir sprechen müssen.
On the whole it wasn't so bad. Nevertheless there were areas of weakness.	Im Großen und Ganzen war es nicht so schlecht. Nichtsdestotrotz gab es Bereiche mit Schwächen.
We've still got some areas to improve.	Wir haben noch ein paar Bereiche zu verbessern.
There were a few mistakes but you really managed quite well.	Da gab es ein paar wenige Fehler, aber mit denen sind Sie ziemlich gut umgegangen.
Apart from the first two months the results have developed really well.	Abgesehen von den ersten zwei Monaten haben sich die Ergebnisse wirklich gut entwickelt.

9.3.2 Protokoll: Mitarbeiterbeurteilung

Template for »Performance Appraisal«			
Employee name		Job title	
Department		Personnel No.	
Reason for review			Annual
	Promotion		Unsatisfactory performance
	End of Probation Period		Other

Date	
employee began present position	
of last appraisal	
Scheduled appraisal date	

Rating identification	
7. - Excellent	Performance is exceptional in all areas and is recognizable as being superior to others.
8. - Very good	Results clearly exceed most position requirements. Performance is of high quality and is achieved on a consistent basis.
9. - Good	Competent and dependable level of performance. Meets performance standards of the job.
10. - Improvement needed	Performance is deficient in certain areas. Improvement is necessary.
11. - Unsatisfactory	Results are generally unacceptable and require immediate improvement. No bonus increase should be granted to individuals with this rating.

General factors	**Rating**	**Supportive details or comments**
Quality: The accuracy, thoroughness, and acceptability of work performed		
Productivity: The quantity and efficiency of work produced in a specified period of time		
Job knowledge: The practical/technical skills and information used on the job		
Reliability: The extent to which an employee can be relied upon regarding task completion and follow-up		

Availability: The extent to which an employee is punctual, observes prescribed work break/meal periods, and the overall attendance record	______________________ ______________________
Independence: The extent of work performed with little or no supervision	______________________

9.4 Vokabeln – Vocabulary input

Nachfolgend finden Sie eine Auflistung wichtiger Fachbegriffe rund um das Thema Referenzen und Arbeitszeugnisse.

English	German
accuracy *Her accuracy leaves a lot to be desired.*	Genauigkeit *Ihre Genauigkeit lässt viel zu wünschen übrig.*
appraisal	Bewertung, Beurteilung
appraisal interview	Mitarbeitergespräch
approach *We need to try a different approach.*	Ansatz *Wir müssen einen anderen Ansatz versuchen.*
attendance *There were 15 employees in attendance.*	Anwesenheit, zugegen sein *15 Mitarbeitende waren zugegen.*
clarify	klären
commitment *I want to thank you for your commitment.*	Einsatz; Engagement *Ich möchte mich für Ihren Einsatz bedanken.*
cope with *I can hardly cope with the workload.*	etwas bewältigen, meistern *Ich kann die Arbeit kaum bewältigen.*
establish	festlegen
fine-tune	fein abstimmen
foster *Our goal is to foster all employees with potential.*	fördern *Unser Ziel ist es alle Arbeitskräfte mit Potenzial zu fördern.*

English	German
goal	Ziel
instruction *The accident happened because the instructions were not clear.*	Anweisung *Der Unfall passierte, weil die Anweisungen nicht klar waren.*
overall	insgesamt
overcome	bewältigen; meistern
peer	Gleichrangige(r)
performance	Leistung
performance agreement	Leistungsvereinbarung
praise	loben
precede: to something ~ *The contract was preceded by long negotiations.*	etwas vorausgehen *Dem Vertrag sind lange Verhandlungen vorausgegangen.*
prescribe	vorschreiben
prioritise	Prioritäten setzen
progress	Fortschritt
promotion *If you do a good job you might get promotion.*	Beförderung *Wenn Sie Ihren Job gut machen, werden Sie eventuell befördert.*
punctual	pünktlich
purpose	Zweck
rating *The company was given a bad rating by the bank.*	Bewertung *Die Firma erhielt eine schlechte Bewertung von der Bank.*
reliability	Verlässlichkeit
review	überprüfen; überdenken
self-assessment	Selbsteinschätzung
sum up	zusammenfassen
supervision	Überwachung
target; achieve one's ~	sein Ziel erreichen
thoroughness	Gründlichkeit
timeline *It is essential that you keep to the timeline.*	Zeitrahmen *Es ist wichtig, dass Sie im Zeitrahmen bleiben.*

English	German
top-down	von oben nach unten
track; get off ~ *Things got off track when we employed him.*	vom Weg abkommen, aus dem Ruder geraten *Die Dinge sind aus dem Ruder geraten, nachdem wir ihn eingestellt hatten.*
unsatisfactory	nicht zufriedenstellend
weakness	Schwäche

9.5 Übungen – Exercise time

Task 1: What are the terms for the following definitions?

1. A person who works with you at the company = ________________
2. Another word for the goal or an objective = ________________
3. The results, good or bad, that an employee achieves in his or her work = ________________
4. The employee who is being evaluated = ________________
5. A list of the items to be discussed = ________________

Task 2: Practise those prepositions.

1. To do my part ________________ achieving something.
2. Award ________________ my individual performance
3. Keep an eye ________________ the goals.
4. Apart ________________ the first two months it was good.
5. Am I holding you back ________________ any way?
6. What ________________ improving yourself as a team member?
7. Can he be relied ________________ to do the job?

Task 3: Which is the correct form?

1. How *successful/successfully* did you meet your goals last year?
2. What was *difference/different* about your job in the last year?
3. A *typically/typical* appraisal interview will include …
4. I've been looking forward *to review/reviewing* recent goals with you.
5. Let's decide on the action to be *done/taken.*
6. We should *hold/keep* in mind that …

ARBEITSMITTEL AUF MYBOOK.HAUFE.DE

Auf der Internetseite zum Buch finden Sie weitere Übungen zu diesem Modul. Darüber hinaus bieten wir Ihnen eine strukturierte Vorlage für die Leistungsvereinbarung mit Ihren Mitarbeitern:innen an. Diese Vorlage können Sie ausdrucken und die Leistungsziele schriftlich festlegen. Zudem erhalten Sie eine Checkliste für das Beurteilungsgespräch.

10 Vergütung

Zusammenfassung

Dieses Modul befasst sich mit der Vergütung von Mitarbeiter:innen. Je nach Situation können Sie zwischen zwei Gesprächsleitfäden wählen:

- Der erste *Gesprächsleitfaden* unterstützt Sie im Gespräch mit einem Bewerber, mit dem Sie über den gesamten Arbeitsvertrag reden.
- Der zweite *Leitfaden* gibt Ihnen Formulierungen für das Gespräch mit Mitarbeiter:innen an die Hand, die eine Gehaltserhöhung fordern. Daran schließt sich ein Beispieldialog an, der zeigt, wie so ein Gespräch aussehen könnte.
- Um das Thema »alternative Vergütungsmöglichkeiten« geht es im zweiten Teil dieses Moduls.

Am Ende des Moduls finden Sie alle wichtigen *Vokabeln* zu diesem Thema und *Übungsaufgaben*, mit denen Sie Ihre Kenntnisse überprüfen können.

Die Meilensteine: Wie Sie die Mitarbeitervergütung verhandeln

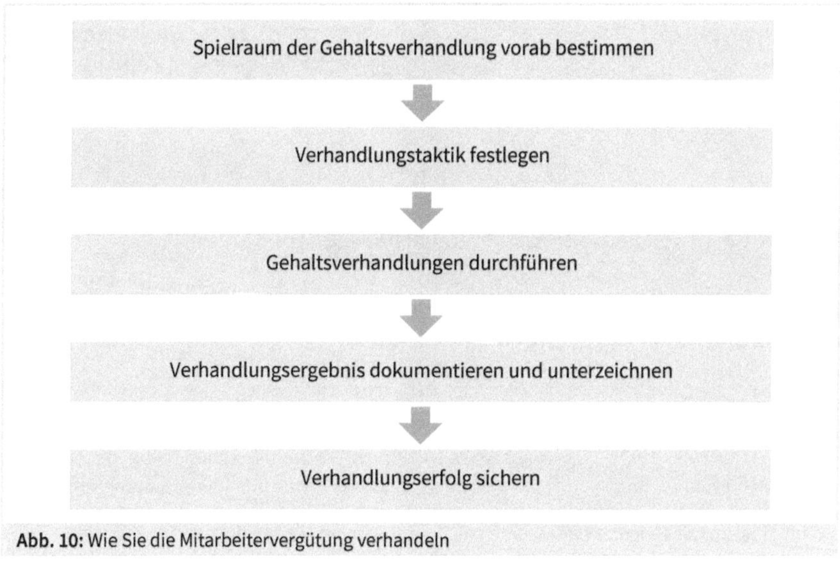

Abb. 10: Wie Sie die Mitarbeitervergütung verhandeln

10.1 Gehaltsverhandlungen führen

Employees often say they would perform better and be better motivated if they had a salary increase. However, there could be other motivators that bring better results such as a language course or the chance to work partly from home.

In any case it is the basic salary increase that plays the main role in most pay negotiations. Nevertheless, it is more sensible both for the employer and the employee to consider the whole salary package. Perks such as private medical insurance, pension and share participation schemes or a company car can all act as a motivating force for the employee.

Mitarbeiter:innen sagen oft, dass sie im Falle einer Gehaltserhöhung bessere Leistungen zeigen würden und auch besser motiviert wären. Obwohl es andere Motivatoren gibt, die zu besseren Ergebnissen führen könnten, wie zum Beispiel Sprachkurse oder die Möglichkeit teilweise von zuhause zu arbeiten. In jedem Falle ist es die Erhöhung der Grundvergütung, die in den meisten Gehaltsverhandlungen die Hauptrolle spielt. Nichtsdestotrotz ist es sowohl für Arbeitgeber als auch für Arbeitnehmer:in sinnvoller, das gesamte Gehaltspaket zu betrachten. Freiwillige Sozialleistungen wie eine private Krankenversicherung, Altersvorsorgen und Mitarbeiterbeteiligungsprogramme oder ein Firmenwagen können sich alle positiv auf die Motivation von Arbeitnehmer:innen auswirken.

10.1.1 Gesprächsleitfaden: Gehaltsverhandlungen führen

Eine weitere Gesprächsvariante zum Thema Gehaltsverhandlungen ist das Gespräch zwischen Personaler und Bewerber:in..

Moving to the topic of salary	Auf das Gehalt zu sprechen kommen
Now as we have dealt with ..., perhaps we can move on to remuneration.	Nun, da wir uns mit ... beschäftigt haben, könnten wir nun über die Vergütung sprechen.
So that's all on this topic. Let's start with pay.	Das ist nun alles zu diesem Thema. Lassen Sie uns mit der Bezahlung beginnen.
We've covered the contract question. Shall we get down to business with salary expectations?	Wir haben die Vertragsfrage geklärt. Sollen wir bezüglich der Gehaltserwartung nun zur Sache kommen?
So contract conditions are done. I'd like to look at the salary structure.	Nun, die Bedingungen des Vertrags sind erledigt. Ich würde gerne auf die Gehaltsstruktur eingehen.

Checking on salary expectations	Nach den Gehaltserwartungen fragen
What sort of salary were you expecting?	Welche Gehaltshöhe haben Sie erwartet?
What are your salary expectations exactly?	Was sind Ihre tatsächlichen Gehaltsvorstellungen?
Well, how much were you expecting to earn approximately?	Nun, wie viel haben Sie ungefähr erwartet zu verdienen?
Do you know anything about our salary structure?	Wissen Sie etwas über unsere Gehaltsstruktur?
What sort of remuneration did/do you have in mind?	Welche Gehaltshöhe schwebt Ihnen vor?
What are your wishes regarding salary?	Was sind Ihre Gehaltsvorstellungen?
The applicant's wishes	**Die Wünsche des Bewerbers**
I was hoping to move up to £ 45,000.	Ich habe gehofft, mich auf 45.000 £ zu verbessern.
I wish to earn at least …	Ich möchte mindestens … verdienen.
I would not want to start below …	Ich möchte nicht unter … anfangen.
I expect something in excess of …	Ich erwarte etwas über …
I'd like to start on at least …	Ich würde gerne mit mindestens … anfangen.
Explaining there are other financial rewards	**Weitere finanzielle Anreize erläutern**
We'd be willing to offer you … but there are other considerations.	Wir wären bereit, Ihnen … anzubieten, aber da sind noch andere Gesichtspunkte.
In addition we have salary increases by negotiation depending on individual projects.	Zusätzlich haben wir verhandelte Gehaltserhöhungen, die von individuellen Projekten abhängig sind.
I ought to point out that we have increases linked to performance targets.	Sie sollten wissen, dass wir leistungsabhängige Gehaltszuwächse haben.
That's the base salary. Then come the bonuses.	Das ist das Grundgehalt. Dazu kommen die Zulagen.
Describing extra benefits	**Sonderleistungen beschreiben**
In addition there are other perks.	Zusätzlich gibt es noch freiwillige Sozialleistungen.
Extra benefits include private health insurance.	Zusatzleistungen schließen eine private Krankenversicherung ein.
Private health insurance comes on top of that.	Die private Krankenversicherung kommt noch dazu.
A special bonus is school fees for private education for your kids.	Eine spezielle Zusatzleistung ist die Übernahme des Schulgeldes für die private Ausbildung Ihrer Kinder.

10.1.2 Gesprächsleitfaden: Vergütung mit neu beförderten Mitarbeitern:innen verhandeln

Natürlich möchte ein Mitarbeiter, der gerade befördert wurde, auch eine entsprechende finanzielle Entlohnung dafür erhalten. Der folgende Leitfaden zeigt den möglichen Aufbau eines solchen Gesprächs zwischen dem:der Personalverantwortlichen und dem:der beförderten Mitarbeiter:in. Sie erhalten Musterformulierungen für folgende Gesprächsabschnitte:

- Einleitung
- Sichtweisen darlegen
- Forderungen identifizieren
- Alternativen aufzeigen
- Was ist das Unternehmen bereit zu zahlen?

Introducing the topic of promotion	**Über Beförderung sprechen**
As you know you have the chance of a …	Wie Sie wissen, haben wir die Möglichkeit für …
You are already aware that we wish to promote you to …	Ihnen ist ja schon bekannt, dass wir Sie zum … befördern möchten.
It's time for you to take on more challenging tasks, including …	Es ist an der Zeit, dass Sie eine weitere Herausforderung annehmen.
Today we have to talk about your promotion.	Heute müssen wir über Ihre Beförderung sprechen.
In this meeting we'd like to talk about your future.	In diesem Meeting würden wir gerne über Ihre Zukunft sprechen.
Introducing an issue in more detail	**Ein Thema detailliert ansprechen**
As you know we've come together to talk about …	Wie Sie wissen, sind wir zusammengekommen, um über … zu sprechen.
Today we need to specifically discuss your …	Heute müssen wir speziell über Ihre … diskutieren.
In this meeting we need to come to an agreement about …	In diesem Meeting müssen wir eine Einigung über … erzielen.
I'd like to go into detail about the ….	Ich würde gerne detailliert auf … eingehen.
So, today it's the salary package. Let's deal with the following items: …	So, heute geht es um das Gehaltspaket. Lassen Sie uns mit folgenden Dingen beschäftigen.
Explaining one's view	**Sichtweise darlegen**
I think it's only fair to get a salary increase considering …	Ich denke angesichts der … ist eine Gehaltserhöhung angemessen.
I've more responsibility. I think that should result in a salary increase.	Ich habe mehr Verantwortung. Ich denke, das sollte eine Gehaltserhöhung nach sich ziehen.

As you have already told me my results are good. I think it's time for a salary improvement.	Wie Sie mir schon gesagt haben, sind meine Arbeitsergebnisse gut. Ich denke, es ist Zeit mein Gehalt zu verbessern.
My situation in the company has changed. That should be reflected in my salary.	Meine Position in der Firma hat sich verändert. Das sollte sich in einer Erhöhung des Gehalts widerspiegeln.
I'm doing much more work. I think the company should recognise this. What can you offer me in that respect? It doesn't necessarily need to be a salary increase.	Ich erledige sehr viel mehr Arbeit. Ich denke die Firma sollte das anerkennen. Was können Sie mir diesbezüglich anbieten? Es muss nicht unbedingt mehr Gehalt sein.
Checking what the employee wants	**Herausfinden was der:die Mitarbeiter:in möchte**
By the way, what sort of salary expectations do you have?	Übrigens, welche Gehaltsvorstellungen haben Sie?
So what were you hoping for in terms of money?	Was haben Sie sich in Bezug auf die Bezahlung vorgestellt?
Well, what sort of benefits do you have in mind?	Nun, an welche Art von Sozialleistungen denken Sie?
I wonder where you'd like to see yourself in five years.	Ich frage mich, wo Sie sich selbst in fünf Jahren sehen.
What do you feel would be fair in the present situation?	Was denken Sie, wäre in der aktuellen Situation angemessen?
Giving alternatives to money increases	**Alternativen zu Gehaltserhöhungen aufzeigen**
A simple salary increase is only one way to …	Eine einfache Gehaltserhöhung ist nur ein Weg um …
I understand your wish but what do you think about other sorts of benefits?	Ich verstehe Ihren Wunsch, aber wie stehen Sie zu anderen Vergünstigungen?
Money is one thing but how about alternatives?	Geld ist eine Sache, aber wie sieht es mit den Alternativen aus?
There are other things, not only salary increases. For example …	Da gibt es andere Dinge außerhalb einer Gehaltserhöhung. Zum Beispiel …
Explaining what the company could offer	**Erläutern, was die Firma bieten könnte**
The company would be in a position to offer profit-sharing …	Die Firma wäre in der Lage, Ihnen eine Gewinnbeteiligung anzubieten.
It might be possible to arrange a company car for you.	Es wäre möglich, Ihnen einen Firmenwagen zu stellen.
We might find ourselves able to …	Wir sind eventuell in der Lage, …

There are several schemes which could fit the bill.	Es gibt mehrere Modelle, die passen könnten.
We could offer you several of the following perks: ...	Wir könnten Ihnen einige der folgenden Sonderleistungen anbieten: ...
Setting up a basis for further negotiation	**Die Grundlage für weitere Verhandlungen legen**
I believe the company would be willing to ...	Ich glaube die Firma wäre bereit zu ...
The company is certainly in a position to ...	Die Firma ist sicherlich in der Lage, um ...
If you can accept ... as a starting point, we could talk about ...	Wenn Sie ... als Ausgangspunkt akzeptieren können, können wir über ... reden.
If you can accept this as a basis, we could talk some more about ...	Wenn Sie dies als Basis akzeptieren, können wir zukünftig über ... reden.

10.1.3 Beispieldialog: Discussing the salary package with a promoted employee

Background: Jim Brims from Human Resources, responsible for personnel development. Jean Joyce, sales manager.

Jim Brims: Well, as you know you have the chance of a small promotion to team leader of a project group in your department.

Jean Joyce: Yes, I've already had a preliminary discussion with the head of department. We're meeting again on Friday.

Jim Brims: As you know we've come together today to talk about your salary package. We've already discussed the main points with your department leader. And today we should come to an agreement concerning the details of the package. Have you anything you'd like to mention now at the beginning?

Jean Joyce: Well, obviously I'd like to talk about a higher salary. I think that isn't unusual considering that I will have a lot more responsibility than before. Before I was alone on the project. In future I'll be in charge of a group of six.

Jim Brims: Yes, we understand that, of course. By the way, what sort of salary expectations do you have in the present situation?

Jean Joyce: I feel that an increment of ten percent gross would be fair for the new situation.

Jim Brims: I see your point, Ms Joyce, However, a simple salary increase is only one way to get fair compensation and to foster employee motivation.

Jean Joyce: Extra money is vital in a country where we have inflation of three percent and hidden costs arising all the time.

Jim Brims: I quite understand. What I'm saying is that there are other ways to approach the topic and get similar results. For example the company would be in a position to offer you a profit-sharing scheme. From our point of view, employees are more motivated to move forward to success, and subsequently earn more. The more profitable the company is, the higher your income.

Jean Joyce: Okay. Well, I hadn't really considered that. But it probably wouldn't bring me the equivalent of ten percent.

Jim Brims: There are other interesting schemes as well. I'll come to them later. But to set your mind at rest I believe the company would be willing to offer you a five percent gross increase. The other perks and schemes would certainly make up the difference to the ten percent. If you can accept that as a starting point then I could describe the extra benefits.

10.2 Vergütungsalternativen erläutern

Clearly the company should try to get the level and scope of the remuneration right for employees so as to support motivation and performance. At the same time alternatives should also be considered.

These demonstrate to employees that their contributions are recognised and respected, and the staff then feel included in the development of the company. Activities which show staff recognition include special events, professional training and equipment (cars, laptop).

Vornehmlich sollte die Firma versuchen, das Niveau und den Rahmen der Entlohnung so zu gestalten, dass dadurch die Motivation und Leistungsbereitschaft der Angestellten unterstützt wird. Gleichzeitig sollten auch Alternativen in Erwägung gezogen werden.
Diese zeigen den Angestellten, dass ihr Engagement bemerkt und respektiert wird. Zudem fühlt sich die Belegschaft in das Geschehen der Firma einbezogen. Maßnahmen, die dem Personal die Anerkennung zeigen, beinhalten spezielle Veranstaltungen, Bildungsmaßnahmen und Ausrüstungen (Fahrzeuge, Laptop).

10.2.1 Gesprächsleitfaden: Vergütungsalternativen erläutern

Nachfolgend finden Sie nützliche Musterformulierungen, um Mitarbeiter:innen alternative Vergütungsmöglichkeiten aufzuzeigen.

Explaining the procedure	**Die Vorgehensweise erläutern**
So first I'd like to explain something about …	So, als erstes würde ich gerne etwas über … erklären.
First of all I want to talk about … and then I'll come to …	Zuallererst möchte ich über … reden und dann werde ich zu … kommen.
I'll begin by …, secondly I'll refer to …, and finally I'll deal with …	Ich werde zuerst mit .. beginnen, dann werde ich mich auf … beziehen und mich schließlich mit … beschäftigen.
So the topic for now is the …	Das Thema für jetzt ist …
Today we have to discuss … in some detail.	Heute müssen wir über … im Detail reden.
Explaining the alternative rewards	**Vergütungsalternativen erläutern**
In addition to base salary and a bonus we have extra benefits/perks.	Außer dem Grundgehalt und einem Bonus haben wir zusätzliche Sozialleistungen.
We can offer several extras. These include …	Wir können einige Extras bieten. Dazu gehören …
Well, you'll be getting a company car.	Nun, Sie werden ein Firmenfahrzeug bekommen.
You'll be able to participate in our share participation scheme and there'll be extra training and coaching opportunities.	Sie können an unserem Aktienbeteiligungsplan partizipieren und es wird noch zusätzliche Angebote zur Weiterbildung geben.
You'll have the chance to improve your qualifications.	Sie werden die Möglichkeit haben, Ihre Qualifikation zu verbessern.
Changing the direction of the discussion	**Zum nächsten Gesprächspunkt wechseln**
Well, let's come to the financial side of things.	Nun, lassen Sie uns zu der finanziellen Seite kommen.
So, perhaps we could talk about the other perks.	Nun, vielleicht könnten wir über die anderen Sozialleistungen reden.
I'd like to put another topic on the table – the question of qualifications.	Ich würde gern einen anderen Punkt thematisieren, und zwar die Frage der Qualifikationen.
To come to another point. What about …?	Um zu einem anderen Punkt zu kommen. Wie sieht es mit … aus?

Adding some extra information	Zusätzliche Informatinen geben
In addition I ought to mention that …	Außerdem sollte ich erwähnen, dass …
Moreover, I wanted to add that …	Darüber hinaus wollte ich hinzufügen, dass …
And there's something else I wanted to say.	Und da gibt es noch etwas anderes, was ich sagen wollte.
There are still a few things I'd like to have done.	Da sind noch ein paar Dinge, die ich gerne erledigt hätte.
I think you might appreciate knowing that …	Ich denke, Sie würden es begrüßen, zu wissen, dass …
Concluding the discussion	**Das Gespräch abschließen**
And finally, you'll receive a training course/ coaching/holiday entitlement of …	Und schließlich erhalten Sie einen Anspruch auf Bildungsurlaub für …
To conclude I'd like to point out that …	Abschließend möchte ich herausstellen, dass …
I'd like to finish by reminding you that …	Ich möchte schließen, indem ich Sie daran erinnere, dass …
So I think we're in agreement finally about …	Nun, ich denke, wir sind uns abschließend darüber einig, dass …
Well, we should call it a day.	Nun, wir sollten für heute Schluss machen.
So I think we've covered everything.	So, ich denke, wir haben alles abgehandelt.

10.2.2 Beispieldialog: HR explains alternative rewards to a new employee

Background: Conversation between Paul Petersen from Human Resources, and Lisa Bond, the new marketing manager.

Paul Petersen: So, in this part of the meeting I'd like to explain something about the company's benefits package and alternative rewards.

Lisa Bond: I'd be very grateful. You hear a lot of rumours but I've not seen anything concrete yet. What's the situation in my particular case?

Paul Petersen: Well, there are quite a few perks to what we call a level 2 position. Shall I explain them one by one as far as your package is concerned?

Lisa Bond: If you would and perhaps I could ask a few questions as we go along?

Paul Petersen: Of course. Please feel free. Well, the first of our extra benefits is the company car. This has tax advantages over simply receiving more salary.

Lisa Bond: Are there any restrictions on the type of car?

Paul Petersen: Yes, indeed there are. We have three levels. Your position qualifies for level two. Typically that's something like an Audi A4 or a VW Golf. And you have to keep a record of your kilometres used for business purposes.

Next perhaps we should talk about insurances. You'll get a private medical insurance with a free medical check-up.

Lisa Bond: Can I choose the company myself or …

Paul Petersen: I'm afraid not. We have an arrangement with P.P.P. That's Private Patient Plan.

Well, let's come to the financial side of things. There's an opportunity for you to participate in our share option scheme.

Lisa Bond: Oh, that sounds interesting. Is there a limit to the number of shares?

Paul Petersen: You can buy one hundred but at sixty percent of the current price. In addition I ought to mention that we have a subsidized canteen offering healthy organic food.

Lisa Bond: Wow! That's a real perk.

Paul Petersen: Finally, you'll receive regular training courses for your personal development. Let me think. Yes, I think that was everything.

10.3 Vokabeln – Vocabulary input

Nachfolgend finden Sie eine Auflistung wichtiger Fachbegriffe rund um das Thema Gehaltsverhandlungen.

English	German
company car	Firmenwagen
compensation *When she was fired she received a high compensation.*	Abfindung *Als sie entlassen wurde, erhielt sie eine hohe Abfindung.*

English	German
contract condition	Vertragsbedingung
contribution	Beitrag
development; personnel ~	Personalentwicklung
excess; in ~ of *We made a profit in excess of EUR 5 million.*	über; höher als *Wir erzielten einen Gewinn von über 5 Millionen EUR.*
expectation	Erwartung
foster	fördern
get down to business	zur Sache kommen
insurance; health ~	Krankenversicherung
insurance; private medical ~	private Krankenversicherung
medical check-up *You need to undergo a medical check-up.*	Kontrolluntersuchung *Sie müssen sich einer Kontrolluntersuchung unterziehen.*
motivators	motivierendes Element
negotiation *The negotiations went very well for the supplier.*	Verhandlung *Die Verhandlungen liefen sehr gut für den Lieferanten.*
negotiations; pay ~	Gehaltsverhandlungen
participate *Last month I participated in a job interview.*	teilnehmen *Letzten Monat nahm ich an einem Bewerbungsgespräch teil.*
performance targets	Leistungsziele
perks	Sonderzulagen
profit-sharing scheme	Gewinnbeteiligungsmodell
promote	befördern
promotion	Beförderung
purpose; business ~	Geschäftszweck
reflect	widerspiegeln
remuneration *Your remuneration will be performance-based.*	Vergütung *Ihre Vergütung ist leistungsabhängig.*
restriction	Einschränkung

English	German
reward *There are alternative rewards such as further education or share participation.*	Entlohnung *Es gibt eine alternative Entlohnung, wie z. B. Weiterbildung oder Aktienbeteiligung.*
rumour	Gerücht
salary expectation	Gehaltserwartung
salary increase	Gehaltserhöhung
salary package *Included in the salary package are children's school fees.*	Gehalt inkl. Sonderleistungen *Zum Gehalt zählen Sonderleistungen wie Schulgebühren für Kinder.*
salary structure	Gehaltsstruktur
salary; basic ~	Grundgehalt
school fees	Schulgebühren
set someone's mind at rest *I would like to set your mind at rest by telling you that the situation has improved.*	jemanden beruhigen *Ich möchte Sie beruhigen, indem ich Ihnen sage, dass sich die Situation verbessert hat.*
share participation scheme	Aktienbeteiligungsplan
staff members	Mitarbeiter; Belegschaft
staff recognition	Anerkennung der Belegschaft
subsidised	subventioniert
tax advantage	steuerlicher Vorteil
topic *The topics of today's meeting include 5 new items.*	Thema *Die Themen der heutigen Besprechung umfassen 5 neue Punkte.*

10.4 Übungen – Exercise time

Task 1: Translate the words in the brackets.

1. The staff need to *(mehr leisten)* if we are going to reach our goals.
1. Perks include a *(private Krankenversicherung)* as well as a company car.
2. I have a lot more experience now. That should be *(widerspiegeln)* in my salary.
3. The employee was *(befördert)* due to a whole series of skills.
4. We have to get the level of *(Bezahlung)* right or the staff will be demotivated.
5. We also offer a *(subventionierte)* cafeteria in the company.

Task 2: Which is the correct word?

1. Staff said they would *motivate/perform* better if they had a pay rise.
2. By the way, what are your salary *wishes/expectations*?
3. The head of R&D did his *promotion/doctorate* at the University of London.
4. Dr. Hohr is the head of *personal/personnel* development.
5. He's in charge of the finance section in John's *compartment/department*.

Task 3: Working on prepositions. Fill in the missing prepositions.

1. So that's all now. Let's move on to remuneration.
2. We have an arrangement P.P.P.
3. I'll explain them one one to you.
4. So, let's come the financial side of things.
5. I'd like to finish explaining ...
6. the same time we should consider alternatives.
7. Jim Brims is responsible personnel development.
8. What were you hoping in terms of salary?

ARBEITSMITTEL AUF MYBOOK.HAUFE.DE

Auf der Internetseite zum Buch finden Sie weitere Übungen zu diesem Modul. Zudem bieten wir Ihnen eine nützliche Checkliste für Gehaltsverhandlungen an. Die Liste behandelt die Punkte: Gehalt, bezahlten Urlaub, Sozialleistungen und andere Leistungen wie zum Beispiel Dienstwagen. Sie können diese Liste ausdrucken und während des Gesprächs einsetzen.

11 Tarifverhandlungen und Gewerkschaften

Zusammenfassung
Dieses Modul befasst sich mit den schwierigen psychologischen Aspekten der Beziehungen zwischen Management und Belegschaft sowie zwischen Arbeitgeberverbänden und Gewerkschaften. Damit Sie solche Situationen erfolgreich meistern, enthält dieses Modul nützliche Tipps und Arbeitsmittel. Am Ende des Moduls finden Sie die für diesen Bereich spezifischen *Fachbegriffe* und *Übungen*, mit denen Sie Ihr Wissen überprüfen können.

Die Meilensteine: Tarifverhandlungen führen

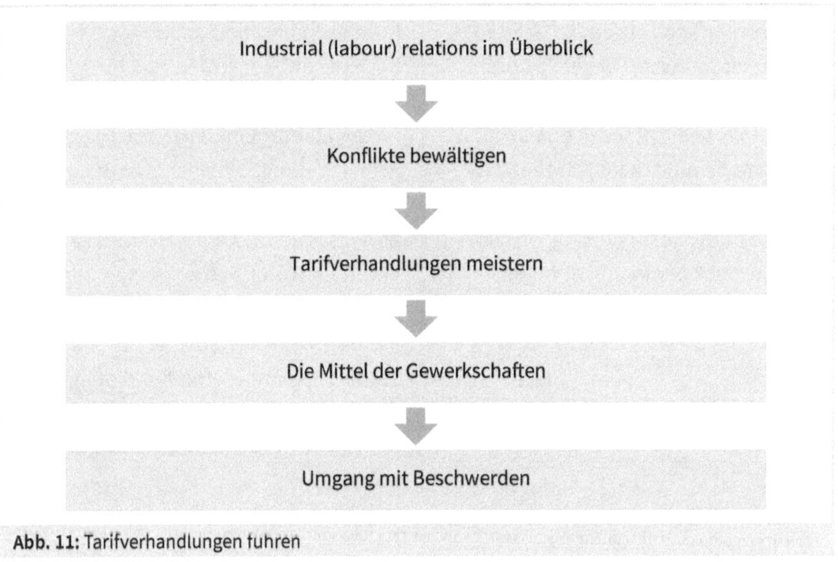

Abb. 11: Tarifverhandlungen führen

11.1 Industrielle Beziehungen – Industrial (labour) relations

A large number of employees still organise themselves in the structure of a trade union. The majority are blue-collar workers although trade unions are becoming increasingly popular also with the white-collar professions such as teaching, medical staffing and local government.
The question is: Why do working people join unions? How are the agreements between employees and employers managed? Research suggests it's not just a matter of getting more money or better conditions. Certainly the unions achieve these goals, but it seems to be more a matter that without the unions workers would not be treated fairly. Fear of losing jobs, low morale, poor communication push workers into the unions.

Eine große Anzahl von Arbeitskräften organisiert sich nach wie vor in gewerkschaftlichen Strukturen. In der Mehrheit handelt es sich um Arbeiter, aber auch bei immer mehr Angestellten, wie Lehrkräften, medizinischem Personal sowie Behörden, werden Gewerkschaften immer beliebter.
Die Frage ist: Warum treten Arbeitnehmer Gewerkschaften bei? Wie werden die Vereinbarungen zwischen Arbeitnehmern:innen und Arbeitgebern gehandhabt? Untersuchungen haben ergeben, dass es nicht nur eine Frage höherer Einkünfte oder besserer Arbeitsbedingungen ist. Sicherlich verfolgen Gewerkschaften diese Ziele, aber es scheint mehr darum zu gehen, dass Arbeitnehmer:innen ohne Gewerkschaften nicht fair behandelt würden. Die Angst, den Arbeitsplatz zu verlieren, geringe Moral und mangelnde Kommunikation treiben Beschäftigte in die Arme der Gewerkschaften.

What leads to conflict?

Rights of employees	Rights of employer
It is still a confusing situation. There is variation in employees' rights between the various states in the USA and between different countries. The European Union is implementing a standardisation of workers' rights across the Union. Trouble begins when an individual or a group feel as if they have been treated unfairly.	Nowadays the employer may give his opinion about the union's activities and the dangers they pose. The employer must be careful not to threaten or promise things or directly interfere with workers' decisions. All these rights become very »rubbery« in the case of loss of business and danger of closure, when people become afraid.

11.1.1 Gesprächsleitfaden: Collective bargaining

Dieser Leitfaden enthält nützliche Formulierungen für Tarifverhandlungsgespräche. Der anschließende Beispieldialog ist ein Gespräch zwischen einem Gewerkschaftsvertreter und der Geschäftsleitung eines Unternehmens.

Starting the conversation	**Gespräch beginnen**
Let's deal with formalities first, …	Lassen Sie uns zuerst mit den Formalitäten beginnen …
Let's take the matter point by point.	Lassen Sie uns die Angelegenheit Punkt für Punkt durchgehen.
I'd like to clarify our position from the beginning/start/outset.	Ich möchte unsere Position von Beginn an klarstellen.
Process	**Prozess**
We'll make the first proposal as a basis for the negotiations.	Wir machen einen ersten Vorschlag als Grundlage für die Verhandlungen.
Let's move on to the first proposal.	Lassen Sie uns zum ersten Vorschlag kommen.
We'd like to deal with the following items.	Wir möchten uns mit folgenden Punkten befassen.
Introducing points	**Punkte vorstellen**
We'll certainly need to talk about …..	Wir müssen ganz sicher über … reden.
The other items include …	Zu den anderen Punkten gehören …
We have to deal with a long list of items.	Wir müssen uns mit einer langen Punkteliste befassen.
Describing industrial action	**Arbeitskampfmaßnahmen beschreiben**
We'll take action.	Wir werden streiken.
Let's down tools!	Lasst uns die Arbeit niederlegen.
We will work-to-rule.	Wir werden Dienst nach Vorschrift machen.
I think we need a go-slow.	Ich glaube, wir brauchen einen Bummelstreik.
Reaction from the employer	**Reaktion des Arbeitgebers**
We'll lock the workers out.	Wir werden die Arbeiter aussperren.
Make sure we get the names of all trouble-causers.	Stellen Sie sicher, dass wir die Namen von allen Störenfrieden haben.

11.1.2 Beispieldialog: Collective bargaining

Background: The representative of the union meets the management representative to negotiate a labor/works agreement. This should be a meeting in good time and carried out in good faith regarding wages, hours and conditions of employment. It can be the negotiation of an agreement or any connected issue and the drafting of a contract is required.

Jim Beam, official negotiator for the Metal Industry, and Hartmut Lang, Head of Human Resources.

Hartmut Lang: To deal with formalities first, I'd like to state that I'm acting here as the officially nominated representative of Busy Printing Presses.

Jim Beam: Thank you, Hartmut. I think we're aware of that. So perhaps we can get down to business.

Hartmut Lang: Okay. So, I hope that both parties are making every reasonable effort to get an agreement today.

Jim Beam: Don't worry! We'll match each of your suggestions with a counter proposal.

Hartmut Lang: I hope you'll have in mind that we don't need any »surface bargaining«, any impossible conditions imposed, no withholding of information or not wanting to negotiate on mandatory items.

Jim Beam: Come on, Hartmut, we don't need any lectures on what we have to do here.

Hartmut Lang: I agree. Then let's move on to item 1 on our agenda today. We would like to include the following items as mandatory: rates of pay, hours of employment, holiday/vacations severance pay and job performance.

Jim Beam: Well, that's an impressive list and I'm sure we'll try to find the right compromises. There are a few items from our side too for this initial agreement meeting.

Hartmut Lang: So fire away!

Jim Beam: Hartmut, we certainly need to talk about things like employee security, union security, and of course the management and union relationship. This hasn't been exemplary in the recent past, as I think you'll agree.

Hartmut Lang: There was no flexibility on your side.

Jim Beam: Maybe. Other points are pensions and pensions of retired employees. And there's the whole question of the international union. And a whole range of issues on industrial relations, strikes and so on.

Hartmut Lang: Looks like a marathon session.

11.2 Die Mittel der Gewerkschaften

Wenn Tarifverhandlungen zwischen Arbeitgebern und Gewerkschaften erfolglos verlaufen und keine Einigung erzielt wird, dann stehen Gewerkschaften eine Reihe von Maßnahmen zur Verfügung, mit denen sie ihren Forderungen Nachdruck verleihen können.

The tools of the unions	
Term	**Definition**
Strike	In general workers refuse to work.
Economic strike	Workers do not agree on a contract and thus cost the company money for wasted time.
Unfair labour practice strike	The workers protest against what they think is illegal conduct by the employer.
Wildcat strike	The strike is not authorised and takes place during the term of a contract.
Sympathy strike	Here a union strike to support another union in an industrial dispute.
Picketing	Carrying of signs showing employees' worries about the activities of the employer.
Corporate campaign	A campaign where the union pushes other unions or organisations within the company to put the company under pressure.
Boycott	Here company employees refuse to buy products/services from their employer.

11.2.1 Gesprächsleitfaden: Arbeitskampfmaßnahmen

Dieser Leitfaden enthält nützliche Formulierungen für Tarifverhandlungsgespräche. Der anschließende Beispieldialog ist ein Gespräch zwischen einem Gewerkschaftsvertreter und der Geschäftsleitung eines Unternehmens.

Objective	Ziel
We need to have a strategy to …	Wir brauchen eine Strategie, um …
It's time to minimise negative effects.	Es ist an der Zeit, negative Effekte zu minimieren.
Let's fix our goals for the next six months.	Lassen Sie uns unsere Ziele für die nächsten sechs Monate festlegen.
Asking for the other side's proposals	**Um Vorschläge der Gegenseite bitten**
Could we hear your suggestions regarding …?	Können wir Ihre Vorschläge zu … hören?
Would you explain to us what you …?	Würden Sie uns erläutern, was Sie …?
Well, what are your proposals on …?	Also, was sind Ihre Vorschläge zu …?
Explaining ideas	**Ideen erläutern**
I'm of the opinion that we should …	Ich bin der Meinung, wir sollten …
My idea is to …	Meine Idee ist, …
I feel we should do the following.	Ich denke, wir sollten Folgendes tun.
Introducing an argument	**Ein Argument vorbringen**
First, I'd like to draw your attention to …	Zuerst, möchte ich Ihre Aufmerksamkeit auf … lenken.
Now we should look at the question of …	Nun sollten wir uns mit der Frage des … befassen.
We think it's vital to …	Wir glauben, dass es von entscheidender Bedeutung ist, dass …
Making a proposal	**Einen Vorschlag unterbreiten**
I'd like to suggest the following.	Ich möchte Folgendes vorschlagen.
We should draw up a plan.	Wir sollten einen Plan aufstellen.
From our point of view we might solve the issue by …	Unserer Meinung nach könnten wir das Problem lösen, indem …
Rejecting a proposal	**Einen Vorschlag ablehnen**
I'm sorry. We can't accept that idea.	Es tut mir leid. Wir können diese Idee nicht annehmen.
We have to say »no« to that idea.	Wir müssen diese Idee verneinen.
We're not in a position to give the green light on that.	Wir sind nicht in der Lage, hierfür grünes Licht zu geben.
Giving an opinion	**Seine Meinung äußern**
In my eyes I think we should …	In meinen Augen sollten wir …
In my view that is …	Meiner Ansicht nach ist dies …
Our position is that we …	Unser Standpunkt ist, dass wir …

11.2.2 Beispieldialog: Preparing the strike

Background: Talks between the steelworker's union and the management of Ohio Steel Plant have broken down. Paul Black, Director of Operations, and Steve Bagel, Head of HR are discussing how they can limit the damage from the imminent strike.

Steve Bagel: We need to have a strategy in place before they go out on strike. We have to minimise the negative effects.

Paul Black: I've been giving some thought to that. I've got a pretty foolproof plan.

Steve Bagel: Let's hear your suggestions on reducing the confusion that it will cause.

Paul Black: We should pay all the striking workers what they are owed by the company on the first day of the strike. That will reduce their motivation to strike.

Steve Bagel: It might do. What about security?

Palu Black: My idea is to bring in extra security workers from outside and even some trailers to house them in. For when things get rough.

Steve Bagel: You're expecting the worst then?

Paul Black: Yes. And in addition management ought to control all the accesses to the premises and we should think about employing some guards to protect good employees going past the picketers.

Steve Bagel: That's hard stuff. Leaving that for the moment, what about the customers?

Paul Black: Yes, we have to notify of them and also prepare an official reaction to all enquiries.

Steve Bagel: Of course! They shouldn't feel inconvenienced by all this trouble.

Paul Black: And let's not forget our suppliers. They will also have to cross the picket line. Maybe we should fix up alternative ways of getting supplies into the company.

Steve Bagel: You're expecting this industrial action to last some time then?

Paul Black: Last time, it was five years ago, it lasted three weeks and cost the company twenty million dollars. So another measure we took at that time was to arrange overnight stays on the premises with food brought in.

Steve Bagel: I've never heard of anything like that.

Paul Black: And following the same line, we should get in touch with the local job centre and put a team of replacement workers in place.

Steve Black: Hmmm! They won't be qualified to perform skilled tasks.

Paul Black: There will be enough experts here to help. Not all the staff is striking. By the way, we have to get all the buildings photographed. The whole facility. Then we can monitor any misconduct. We need to record any strikers' activities, any violence or threats, property damage and so on. We should collect and document all the data about the strikers and pickets for later use.

11.3 Umgang mit Beschwerden am Arbeitsplatz

Grievance procedure is difficult because some laws are open to interpretation. Employer and the union attempt to decide if the employer indeed violated the contract.	Die Vorgehensweise bei eingehenden Beschwerden ist schwierig, denn nicht immer ist die Rechtslage klar. Arbeitgeber und Gewerkschaft versuchen zu entscheiden, ob in der Tat eine Vertragsverletzung seitens des Arbeitgebers vorliegt.
Grievances very often involve disputes about cases of discipline, redundancies, overtime, vacation and work assignments etc.	Bei den Beschwerden geht es häufig um Disziplinverstöße, betriebsbedingte Kündigungen, Überstunden, Urlaub und Arbeitsaufgaben etc.
Typical examples of grievances:	Typische Beispiele für Beschwerden sind:
Insubordination → Company fired the employee because on two occasions he or she failed to attend an appointment with the boss. He or she filed a complaint against the company.	**Insubordination** → Ein Unternehmen hat einen:e Mitarbeiter:in entlassen, da er bzw sie zwei Termine mit einer Führungsperson versäumt hat. Er:Sie hat eine Beschwerde gegen das Unternehmen eingereicht.

Absenteeism → The company fires an employee for being away from work too often. This employee complained that he or she was not warned or disciplined.	**Abwesenheit** → Ein Unternehmen entlässt einen:e Mitarbeiter:in aufgrund zu vieler Fehlzeiten am Arbeitsplatz. Dieser:e Mitarbeiter:in hat sich beschwert, dass er bzw. sie nicht gewarnt oder diszipliniert wurde.
Plant rules → A rule was implemented that there should be no eating or drinking between breaks. The employee filed a grievance stating there was no reason for such a rule.	**Werksvorschriften** → Es wurde eine Vorschrift eingeführt, dass zwischen den Pausen nicht gegessen oder getrunken werden darf. Der:Die Mitarbeiter:in hat eine Beschwerde eingereicht mit dem Inhalt, dass es keinen Grund für eine solche Vorschrift gibt.
Overtime → In this case the company stopped Sunday overtime for company reasons. Workers filed a grievance for loss of earnings.	**Überstunden** → In diesem Fall hat das Unternehmen aus betrieblichen Gründen Überstunden an Sonntagen gestrichen. Die Belegschaft hat Beschwerde wegen Einkommensverlusten eingereicht.

11.3.1 Mit Beschwerden von Mitarbeitern:innen umgehen

Arbeitnehmer:innen haben das Recht, sich zu beschweren. Wie Sie am besten damit umgehen und was Sie besser unterlassen sollten, erfahren Sie hier.

Tips for dealing with grievances	
What the company should do	**What the company should not do**
• Treat each other fairly as it could come to arbitration. • Talk to the employee, give him a chance to express himself. • Demand that the union specify which provisions in the contract have been violated. • Follow the rules for handling grievances. • Check out the work environment of the grievance. • Check the grievant's record. • Consider the union as an equal. • Inform the supervisor about the grievance.	• Discuss the case with a union representative alone. • Organise something with individual employees which are at odds with the contract. • Give away your rights as a manager to the union. • Solve the grievance based on what you think is fair. You must comply with the contract rules. • Pretend that grievances don't exist because your »managerial« hands are tied. • Make informal changes to the contract.

11.4 Vokabeln – Vocabulary input

Nachfolgend finden Sie eine alphabetische Auflistung wichtiger Fachbegriffe und Beispielsätze zum Thema:

English	German
absenteeism	Abwesenheit; Fehlzeiten
access; get ~	Zutritt erhalten
agreement *We signed an agreement with the works council.*	Vereinbarung *Wir haben eine Vereinbarung mit dem Betriebsrat unterzeichnet.*
arbitrate	schlichten
arbitration	Schlichtung
blue-collar worker	Arbeiter
boycott	Boykott
break off talks *The talks were broken off with no results.*	Gespräche abbrechen *Die Gespräche wurden ergebnislos abgebrochen.*
closed shop	Betriebe, in denen die Beschäftigten einer Gewerkschaft angehören müssen
collective bargaining	Tarifverhandlungen
compromise *In the end a satisfactory compromise was found.*	Kompromiss *Letztlich wurde ein zufriedenstellender Kompromiss gefunden.*
condition	Bedingung
conditions of employment	Arbeitsbedingungen
cross the picket line	Streikposten durchbrechen
effort; make an ~	sich bemühen
employee security	Mitarbeitersicherheit
employees' rights	Arbeitnehmerrechte
expect the worst	vom Schlimmsten ausgehen
feel inconvenienced	sich belästigt fühlen
follow the rules	Regeln befolgen
formality *There were a lot of formalities to deal with.*	Formalität *Es gab eine Menge Formalitäten zu erledigen.*

English	German
get down to business	eine Aufgabe anpacken
grievance	Beschwerde
grievance procedure	Beschwerdeverfahren
hold talks *It is necessary that the union holds talks with the works council first.*	Gespräche führen *Es ist notwendig, dass die Gewerkschaft zuerst mit dem Betriebsrat Gespräche führt.*
impose *The company imposed new restrictions.*	auferlegen *Das Unternehmen hat neue Einschränkungen auferlegt.*
insubordination	Befehlsverweigerung
job performance	Arbeitsleistung
last *The strike lasted several months.*	dauern; anhalten *Der Streik hielt über mehrere Monate an.*
mandatory *It is mandatory for employees to take a lunch break.*	zwingend vorgeschrieben *Mittagspausen sind für Beschäftigte zwingend vorgeschrieben.*
negotiate *We negotiated a fair deal.*	verhandeln; aushandeln *Wir haben einen fairen Deal ausgehandelt.*
open shop	Betriebe, in denen es den Beschäftigten freigestellt ist, ob sie Mitglied einer Gewerkschaft sein möchten.
overtime *The agreement specifies that workers must not do more than 20 hours overtime per months.*	Überstunden *Die Vereinbarung legt fest, dass Arbeitskräfte nicht mehr als 20 Überstunden pro Monat leisten dürfen.*
picketing	Streikposten aufstellen
pose a danger	eine Gefahr darstellen
profession	Beruf
proposal *The proposal was accepted by all parties.*	Vorschlag *Der Vorschlag wurde von allen Parteien angenommen.*
rate of pay	Lohnsatz
records; to take ~	Buch führen; Aufzeichnungen machen
reminder *This is the second reminder we have sent to this customer.*	Mahnung *Dies ist die zweite Mahnung, die wir an diesen Kunden geschickt haben.*

English	German
retire *He was forced to retire early.*	in Rente gehen *Er wurde gezwungen, früher in Rente zu gehen.*
severance pay	Abfindungszahlung
strike action	Arbeitskampfmaßnahme
strike; go on ~	streiken
subordinate	untergeordnet; nachrangig; Befehlsempfänger
threaten someone	jemandendrohen
trade union	Gewerkschaft
treat fairly	gerecht behandeln
union member	Gewerkschaftsmitglied
union representative	Gewerkschaftsvertreter
union; join a ~	einer Gewerkschaft beitreten
vacation; take ~	Urlaub nehmen
violate *The employees violated the contract by going on strike.*	verletzen *Die Beschäftigten haben den Vertrag verletzt, indem sie streikten.*
white-collar worker	Angestellter

11.5 Übungen – Exercise time

Task 1: Put the words in the box into the sentence in the correct form.

signify	achievement	secure	structure

1. We are losing money. The company will have to be ________________.
2. Our data bases have a high level of ________________.
3. With this approach we'll never ________________ a win-win situation.
4. What is the real ________________ of these figures?

Task 2: Give the opposite for:

1. unionist = ________________
2. withdraw from = ________________
3. dismiss = ________________
4. fairly = ________________
5. worse conditions = ________________

Task 3: What is the English for the following phrases?

1. Lassen Sie uns die Angelegenheit Punkt für Punkt besprechen.
2. Sehr interessant, aber wie möchten noch folgenden Punkt mitaufnehmen.
3. Lasst uns die Arbeit niederlegen!
4. Kümmern wir uns zuerst um die Formalitäten

ARBEITSMITTEL AUF MYBOOK.HAUFE.DE

Auf der Internetseite zum Buch finden Sie weitere Übungen zu diesem Modul sowie Gesprächsleitfäden und zahlreiche Textbausteine.

12 Mitarbeitermotivation

Zusammenfassung
In diesem Modul zeigen wir Ihnen, wie Sie die Motivation im Team erhalten und steigern können. Im *Gesprächsleitfaden* finden Sie nützliche Redemittel rund um Themen wie Leistung und Motivation..

Am Ende dieses Moduls finden Sie alle wichtigen themenspezifischen *Vokabeln* und drei *Übungen*, mit denen Sie Ihr Wissen testen können.

Die Meilensteine: Mitarbeiter motivieren

Abb. 12: Mitarbeiter:innen motivieren

12.1 Demotivierende Führungsfehler erkennen

The leader of a group of employees has to make certain that the team works hard and effectively. Staff should be willing to offer extra time and effort in special circumstances. The manager should ensure that the job is interesting, challenging and worthwhile.

Eine Führungskraft, die ein Team leitet, muss sicherstellen, dass das Team hart und effektiv arbeitet. Das Personal sollte unter besonderen Umständen bereit für Überstunden sein und zusätzliche Einsatzbereitschaft zeigen. Die Führungskraft sollte sicherzustellen, dass die Arbeit interessant, herausfordernd und lohnend ist.

Supervisors need to be certain that all team members know what is expected of them and that recognition is directly connected to results and effort.

Vorgesetzte müssen sich sicher sein, dass alle im Team wissen, was von Ihnen erwartet wird, und dass Anerkennung in direktem Zusammenhang mit Ergebnissen und Einsatz steht.

12.1.1 Gesprächsleitfaden: Gründe für einen Leistungsabfall im Team erörtern

Der folgende Leitfaden enthält Musterformulierungen für Gespräche zwischen Personalabteilung und Führungskraft. Er dient dazu, die Motivierungspotenziale im Team zu identifizieren.

Introducing a potential problem	**Ein mögliches Problem ansprechen**
Perhaps it's something in the day-to-day running of things.	Vielleicht liegt es an der täglichen Routine.
It could be a matter of the daily activities in the office.	Es könnte eine Sache der täglichen Aktivitäten im Büro sein.
You could possibly be too close to the everyday work.	Möglicherweise sind Sie zu nah an der täglichen Arbeit dran.
Maybe it's really a question of …	Vielleicht ist es wirklich eine Frage von …
Suggesting what might be the problem	**Andeuten, was das Problem sein können**
It could be that employees feel they are not trusted/valued.	Möglicherweise fühlen Mitarbeiter , dass man ihnen nicht vertraut/sie nicht schätzt.
Employees might possibly feel misunderstood.	Die Angestellten fühlen sich möglicherweise nicht verstanden.

Perhaps staff might somehow feel they are being watched.	Vielleicht fühlt sich die Belegschaft irgendwie beobachtet.
Maybe the staff think Big Brother is watching them.	Vielleicht denkt die Belegschaft, dass »Big Brother« sie beobachtet.
Suggesting ideas	**Ideen vorschlagen**
Let's talk about at a few ideas.	Lassen Sie uns über ein paar Ideen reden.
Shall we check out a few ideas?	Sollen wir ein paar Ideen überprüfen?
What about brainstorming?	Was halten Sie davon, wenn wir zusammen Ideen sammeln?
Could we think about a few points?	Könnten wir über ein paar Punkte nachdenken?
Introducing a list of items	**Eine Reihe von Punkten einbringen**
First let's talk about the following ...	Lassen Sie uns zuerst über Folgendes sprechen.
I'd like to begin by going through several items.	Ich möchte damit beginnen, ein paar Punkte durchzugehen.
Firstly we can deal with ..., then we could move on to ...	Zuerst können wir uns mit ... beschäftigen und dann zu ... übergehen.
There are several things we should talk about. For example, ...	Es gibt einige Dinge, über die wir reden sollten. Zum Beispiel ...
Suggesting additional ideas	**Zusätzliche Ideen vorschlagen**
There are other important things you could do.	Es gibt andere wichtige Dinge, die Sie tun könnten.
If I might suggest another idea here.	Wenn ich an dieser Stelle eine andere Idee vorschlagen dürfte.
By the way, there are some other things we should think about.	Nebenbei bemerkt, es gibt einige andere Dinge, über die wir nachdenken sollten.
What about another idea? We could ...	Wie sieht es mit einer anderen Idee aus? Wir könnten ...
Introducing an (unwanted) idea	**Eine (unerwünschte) Idee einbringen**
If I might be allowed a comment.	Wenn Sie mir einen Kommentar gestatten.
You probably don't want to hear it, but ...	Sie möchten das wahrscheinlich nicht hören, aber ...
You won't like the sound of it, but ...	Sie werden das nicht hören wollen, aber ...
I know you're not exactly in favour of it, but ...	Ich weiß; Sie sind nicht gerade dafür, aber ...

12.1.2 Beispieldialog: How can I motivate the staff?

Background: HR Consultant, Jefferson Morgan, is talking to the Managing Director of a medium-sized company, Ian McDonald.

Jefferson Morgan: So, Mr Bolt, if I've understood you correctly, your salary arrangements are more than adequate. Your bonus system works, there are extras, but still you're worried about negative thinking and lack of motivation amongst your employees. Is that right?

Ian McDonald: Yes, that's right. I can't quite understand why that should be, but I have that feeling.

Jefferson Morgan: Okay. Perhaps it's something in the day-to-day running of things. Are you personally involved? I mean are you present amongst the employees?

Ian McDonald: Well, I try to be there as often as I can.

Jefferson Morgan: Of course. On the other hand, perhaps staff might somehow feel they are being watched, you know, monitored. That doesn't support the idea of trust too much.

Ian McDonald: It certainly isn't meant like that.

Jefferson Morgan: I'm sure it's not. But it's a matter of how the individuals perceive it. It's nothing we can't adjust. It's a small change in approach. Let's look at a few ideas, if you're happy to do that.

Ian McDonald: I don't know if it'll help but ...

Jefferson Morgan: First let's talk about flexibility. What are your hours of work?

Ian McDonald: We're a nine-to-five firm. They all understand that.

Jefferson Morgan: Do they? Maybe they don't. If you allow more flexible hours, they'll have more family life. They're more motivated. Better results! Then there's the personal touch. If an employee does really good work, you could recognise it with a handwritten letter to him. He'll get the message that you have confidence in his work.

Ian McDonald: Well, I never thought of that.

Jefferson Morgan: There are other small but important things you as the boss could do. For example, think about their health. Bring fruit into the office, a few dishes here and there. It makes a difference and shows you care about their well-being.

Ian McDonald: Well, that's simple enough.

Jefferson Morgan: What about the atmosphere in the office?

Ian McDonald: They work away quietly enough. It's okay.

Jefferson Morgan: If I might be allowed a comment, silent working is not always efficient working. Maybe we could try another approach. Try allowing them to listen to music as long as it doesn't disturb. Socialising amongst the staff in general can raise the real level of performance.

12.2 Motivierend kommunizieren im Team

The atmosphere and mood in the office is very important for achieving good results as a whole. Below you will find some tips on what to do to stimulate motivation among colleagues.

Die Atmosphäre und Stimmung in einem Büro sind insgesamt sehr wichtig, um gute Ergebnisse zu erreichen. Nachfolgend finden Sie einige Hinweise, was Sie tun sollten, um die Motivation unter Kollegen:innen zu fördern.

Dos and Don'ts

+	**Dos**
Good example, produtive and cheerful.	
Support communication amongst staff.	
Be open, professional and friendly.	
Set up a positive atmosphere.	
Keep an open ear – Listen to what they say.	
-	**Don'ts**
Don't be biased or prejudiced about staff. Look at facts!	
Don't be wishy-washy about the goals of the job. Staff should understand the whole picture.	
Don't worry and complain about trivia. The priority is for staff to deliver excellent work.	

12.2.1 Gesprächsleitfaden: Motivierend kommunizieren

Im folgenden Leitfaden finden Sie eine Reihe nützlicher Phrasen, mit denen Sie die interne Kommunikation verbessern und damit die Motivation steigern können.

Being cheerful	**Aufmunternde Worte**
I know it's a lot of work, but we're all doing well.	Ich weiß, das ist eine Menge Arbeit, aber wir werden das schon schaffen.
I'm tired and I'm sure you are too, but we're on the right track.	Ich bin müde und ich bin sicher, Sie auch, aber wir sind auf dem richtigen Weg.
You're doing a great job, but don't let it get you down.	Sie machen eine großartige Arbeit, aber lassen Sie sich von der Arbeit nicht unterkriegen.
Being supportive	**Unterstützende Worte**
You seem to be a bit isolated/on your own. How could we improve things?	Es scheint so, als wären Sie ein wenig isoliert/auf sich allein gestellt. Wie können wir die Dinge verbessern?
Have you any idea how we could make this more user-friendly?	Haben Sie irgendeine Idee, wie wir das anwenderfreundlicher machen könnten?
What could we do to make your work here easier/more fulfilling?	Was können wir tun, um Ihre Arbeit hier leichter/erfüllender zu gestalten?
Being open and friendly	**Offene und freundliche Worte**
Ms Klein, I wanted to say I think you're managing this project really well.	Frau Klein, ich möchte Ihnen sagen, dass Sie das Projekt wirklich gut managen.
You're doing a fine job.	Sie machen eine gute Arbeit.
I know you were a bit nervous this morning, but it's absolutely understandable.	Ich weiß, Sie waren heute Morgen ein bisschen nervös, aber das ist absolut verständlich.
Setting up a positive atmosphere	**Eine positive Atmosphäre schaffen**
It's been hard, I know, but I really appreciate your effort.	Ich weiß, es war hart, aber ich schätze Ihre Anstrengungen wirklich sehr.
It doesn't always run smoothly but it's great the way we pull together.	Es läuft nicht immer alles glatt, aber es ist großartig, wie wir das hin bekommen haben.
I hope you all understand how grateful the company is.	Ich hoffe, Sie alle verstehen, wie dankbar das Unternehmen ist.
Listening to what others say	**Hören, was andere zu sagen haben**
That sounds like a very interesting idea. Could you tell me more?	Das klingt nach einer sehr interessanten Idee. Können Sie mir mehr darüber sagen?
I'm sorry. I didn't catch that. Could you tell me again?	Tut mir leid. Ich habe das nicht mitbekommen. Könnten Sie mir das nochmal sagen?
Okay. I know what you're saying. Could we go into detail?	Okay. Ich weiß, was Sie meinen. Können wir ins Detail gehen?

12.2.2 Beispielrede: Motivating the staff

Ladies and Gentlemen,

I am honoured to reply to the toast of the prosperity of the company, which has been so generously proposed and so genuinely received. This reality alone is, I think, the safest assurance of our future well-being. The success of every organisation, commercial or otherwise, depends primarily on the co-operation of those who run it, and there is no substitute for good relations and teamwork. The same is true for our company.

So when I thank you for the toast I must thank you also for all that you have done for the company in the last twelve months. I hope you will always think of this ... *(Pause)* ... ship not as »it«, but as »us«. The fact is that every one of us here tonight has a real stake in it. *(Pause)*

I am not going to talk shop on an occasion like this, but I do want to remind you once again of a matter of internal policy which is very dear to our hearts. Each member of the team has his or her own job to do, and although we should interact as far as possible, our activities are too complicated to allow us to play musical chairs. But we do want to stimulate your motivation, not only for your own particular job, but for the work of our company as a whole; and more than anything else we want your ideas and suggestions for raising our general efficiency even above its present high level. I believe that there is nothing so destructive to motivation and enthusiasm as a feeling that initiative is not wanted. I assure you that it is desired and needed here, and that the ideas of the most junior members of the staff are not only really welcomed but will always be treated with the consideration and respect they deserve.

That brings me back to teamwork. I think a good company should be like a good cricket team – all working together for the common good, but aiming at quality and equality of performance at the highest level. We are all, still, individuals with our own particular abilities, and I think that individuality should always be supported within the framework of the team.

Finally, I have pleasure in announcing that as from today our bonus scheme will be restructured in the following way. Every employee will participate in the company's shares. There will be a Christmas bonus so that in total every employee will receive thirteen and a half salaries. Finally, there will be a success bonus fixed as a percentage. You may also be pleased to hear that the Woks restaurant will be enlarged. On behalf of the management, Ladies and Gentleman, I thank you.

12.3 Mitarbeitermotivation

The idea that an increase in salary will motivate employees to achieve the agreed targets is widespread. In fact the motivation created through financial incentives soon loses its power.

Weit verbreitet ist die Vorstellung, dass eine höhere Vergütung Mitarbeitende motivieren würde, die vereinbarten Ziele zu erreichen. Tatsächlich verliert jedoch eine Motivation, die durch finanzielle Anreize erzeugt wurde, recht schnell ihre Kraft.

However, it is much more effective to »allow« or even encourage the existing motivation of employees.

Viel wirksamer ist es dagegen, die vorhandene Motivation der Mitarbeitenden zu »erlauben« oder sogar zu stärken.

For example, you can entrust employees who find it easy to establish contacts with other people with tasks where it is important to work together with other people.

So können Mitarbeiternde, denen es leicht fällt, Kontakte zu anderen Menschen zu knüpfen, zum Beispiel mit Aufgaben betraut werden, in denen es darauf ankommt, mit anderen Menschen zusammenzuarbeiten.

Employees who like to perform can be supported in their motivation by giving them the task of developing new objectives or improving the quality of products.

Mitarbeiter, die gerne Leistungen erbringen, können dadurch in ihrer Motivation unterstützt werden, indem Sie Ihnen zum Beispiel die Aufgabe übertragen, neue Ziele zu entwickeln oder die Produktqualität zu verbessern.

And you can motivate employees who strive for more power by checking whether and how you can delegate more decision-making authority and responsibilities to them.

Und Mitarbeiter, die nach mehr Macht streben, können Sie motivieren, indem Sie prüfen, ob und wie Sie Ihnen mehr Entscheidungsfreiheit und Verantwortung delegieren können.

In addition the following question is important: What can we do to stimulate the environment and keep it positive? There are a series of actions and approaches we can take. The basic principle is that the staff should feel that management has a genuine interest in them as people.

Darüber hinaus ist die folgende Frage wichtig: Wie können wir das Arbeitsumfeld fördern und positiver gestalten? Es gibt eine Anzahl von Möglichkeiten und Herangehensweisen, die wir einschlagen können. Das Grundprinzip ist, dass die Belegschaft den Eindruck haben sollte, dass das Management an ihnen als Menschen interessiert ist.

12.3.1 Gesprächsleitfaden: Motivationen erkennen und fördern

Der folgende »Step-by-Step Guide« enthält Musterformulierungen, die nach den verschiedenen Möglichkeiten zur Mitarbeitermotivation gegliedert sind.

Building a career	**Eine Karriere aufbauen**
Perhaps there's some way we can support you in your career targets.	Vielleicht gibt es eine Möglichkeit, wie wir Sie bei Ihren Karrierezielen unterstützen können.
What do you like about your present job and how do you see the future?	Was mögen Sie an Ihrer gegenwärtigen Aufgabe und wie sehen Sie die Zukunft?
What do you feel works best right now?	Was klappt Ihrer Meinung nach am besten?
What do you think your strengths and weaknesses are regarding the future?	Wo sehen Sie Ihre Stärken und Schwächen in der Zukunft?
How do you see your future activities here?	Wie sehen Sie Ihre zukünftigen Aktivitäten hier?
Social gatherings	**Feiern, Geselliges Beisammensein**
You and the company should celebrate birthdays and anniversaries with cards, gifts and flowers. Families should be invited to events and company parties.	Sie und die Firma sollten Geburtstage und Jubiläen mit Glückwunschkarten, Geschenken und Blumen feiern. Familienmitglieder sollten zu Veranstaltungen und Betriebsfeiern eingeladen werden.
Respecting the family	**Die Familie achten**
I believe we should keep track of births, birthdays and anniversaries.	Ich denke, wir sollten die Geburten, Geburtstage und Jubiläen nicht aus den Augen verlieren.
Listen, people, what about a company event to …?	Hört zu Leute, was haltet Ihr von einer Firmenveranstaltung zu …?
I'd like to suggest a summer/Christmas/champagne party.	Ich würde gerne eine Sommer-/Weihnachts-/Champagner-Party vorschlagen.
Please bring your families to our Open Day on …	Bringen Sie bitte Ihre Familien zu unserem Tag der offenen Tür am … mit.
Working at home	**Zu Hause arbeiten**
So you feel you could occasionally do this work at home?	Sie denken also, Sie könnten diese Arbeit hin und wieder von zuhause aus erledigen?
If I've understood correctly you're interested in working from home?	Wenn ich richtig verstanden habe, sind Sie an einem Heimarbeitsplatz interessiert?
We've thought about your idea of working part-time from home. We're open to suggestions.	Wir haben über Ihre Idee eines Teilzeitheimarbeitsplatzes nachgedacht. Wir sind offen für Vorschläge.
It shouldn't be a problem if you to deal with some of your work from home.	Es sollte kein Problem sein, einen Teil der Arbeit von Ihrem Zuhause aus zu erledigen.

Training as a motivator	Weiterbildung als Motivator
How do you feel you would benefit from a training course?	Glauben Sie, dass Sie von einer Schulung profitieren würden?
Which skills do you think you should focus on?	Auf welche Fähigkeiten, denken Sie, sollten Sie sich konzentrieren?
What sort of training programme would support you most?	Welche Art von Schulungsmaßnahme würde Ihnen am meisten helfen?
Where might you have weaknesses that we could support with training/coaching?	Wo haben Sie möglicherweise Schwächen, bei denen wir Ihnen mit Schulungs-/Coaching-maßnahmen helfen könnten?
Saying thank you	**Sich bedanken**
I wanted to thank you for all your effort and support.	Ich möchte mich für Ihren Einsatz und Ihre Unterstützung bedanken.
In case I forget, thank you for your willingness to …	Für den Fall, dass ich es vergesse, möchte ich Ihnen für Ihre Bereitschaft zu … danken.
You've done a great job.	Sie haben großartige Arbeit geleistet …
Actually, you couldn't have done more.	Sie hätten wirklich nicht mehr tun können.
Thank you. That was one hundred percent effort.	Ich danke Ihnen. Das war ein hundertprozentiger Einsatz.
I'm grateful for your contribution.	Ich bin dankbar für Ihren Beitrag.

12.3.2 Beispielgespräch: Dialogue on motivating the staff

Background: This is a dialogue between the personnel manager Tom Lee talking to the head of purchasing Bob Sanchez about some ideas on motivating staff.

Tom Lee: So it's my feeling that we have to try a slightly different strategy to push a really positive feeling in the group. For example, we should keep track of birthdays, anniversaries and new births.

Bob Sanchez: And offer some sort of congratulations, you mean.

Tom Lee: Exactly. We can discuss details later. Then perhaps management should show a bit of flexibility. I heard that several team members have been asking about the possibility of working from home.

Bob Sanchez: That's right. Les and Phil mentioned that to me a couple of days ago.

Tom Lee: We should talk to them about it. It shouldn't be a problem. Not for everybody but a couple of the jobs could be partly covered that way.

Bob Sanchez: That will make their day, I guess. And there's another issue. Some people need some training for the job. We could parcel that up nicely as a motivator. It shows the company cares about the employees' development.

Tom Lee: There is another thing too which affects you.

Bob Sanchez: Oh and what's that?

Tom Lee: Well, you're the boss of the department and I guess some words of thanks or praise would be appreciated.

Bob Sanchez: I do discuss these things in the annual appraisal.

Tom Lee: I know, but perhaps you should say a few words on the spot, so-to-speak.

Bob Sanchez: You could be right. And I was also thinking that I should give the staff a pep-talk about their careers from time to time.

Test: The ten golden questions on motivation

Beantworten Sie die folgenden 10 Fragen und erfahren Sie etwas über Ihre Motivationsqualitäten.

Test: Golden questions on motivation			
		Yes	No
1.	Do you get to know your staff individually? Do you know what is going on in their lives?		
2.	Do you now what interests them at work and outside?		
3.	Do you give them new challenges? Do you realize when they have reached their limits?		
4.	Do you try to help them in their problems with other parts of the company?		
5.	Are you able to share information with them? They'll get hold of it anyway.		
6.	Do you encourage staff to offer new ideas on how things can be improved?		

Test: Golden questions on motivation			
7.	Are you able to delegate the interesting parts of your own work? It's a great incentive to motivation.		
8.	Do you praise immediately for a job well done and not just later in the annual performance assessment?		
9.	Do you spread the word in the company when your people do well?		
10.	Do you involve the staff in at least discussions, if not decisions, on topics that will affect their future?		

Assessment

If you ticked »yes« 4 times or less you should work on yourself.

If you ticked »yes« 5-7 times perhaps something could be improved.

If you ticked »yes« 8-10 times you're a sensitive leader and on the right track.

12.4 Vokabeln – Vocabulary input

Nachfolgend finden Sie eine Auflistung wichtiger Fachbegriffe für die Mitarbeitermotivation .

English	German
adequate *It was an adequate result for the year.*	angemessen *Es war ein angemessenes Ergebnis für dieses Jahr.*
announce *It is my pleasure to announce the winners.*	bekanntgeben *Es ist mir ein Vergnügen, die Gewinner bekanntzugeben.*
annual *Everybody must attend an annual appraisal interview.*	jährlich *Jeder muss an einem jährlichen Mitarbeitergespräch teilnehmen.*
assurance	Gewissheit; Sicherheit
biased	voreingenommen
celebrate	feiern
challenging *It is a challenging task that lies ahead of us.*	herausfordernd *Es ist eine herausfordernde Aufgabe, die vor uns liegt.*

English	German
check something	etwas überprüfen
cheerful	fröhlich, heiter
comment	Kommentar
contribution	Beitrag
day-to-day *Day-to-day tasks can be quite boring.*	täglich *Tägliche Aufgaben können ziemlich langweilig sein.*
delegate	delegieren
favour; in ~ of	für etwas sein
framework; within the ~ of	im Rahmen von
grateful *I'm grateful for your help.*	dankbar *Ich bin Ihnen für Ihre Hilfe dankbar.*
honoured	geehrt
incentive	Anreiz
isolated *I felt very isolated.*	isoliert; abgegrenzt *Ich habe mich sehr ausgegrenzt gefühlt.*
mention	erwähnen
mood *There is a bad mood amongst the staff.*	Stimmung; Laune *Es herrscht eine schlechte Stimmung unter den Mitarbeitern.*
performance assessment	Leistungsbewertung
prejudice *The report ignores the facts and is full of prejudice.*	befangen sein; Vorurteile haben *Der Bericht ignoriert die Tatsachen und ist voller Vorurteile.*
remuneration	Vergütung
reward *She received a reward for long years of service.*	Anerkennung; Belohnung *Sie erhielt eine Anerkennung für die langjährige Dienstzeit.*
salary arrangement	Gehaltsvereinbarung
socialise	Beziehung pflegen
stake *We all have a stake in the firm. It's in all our interests.*	Anteil *Wir alle haben einen Anteil am Unternehmen. Es ist in unser aller Interesse.*
track; keep ~ of	Überblick behalten

English	German
trivia	Belanglosigkeiten
trusted *She is a trusted colleague who all the staff can rely on.*	vertraut *Sie ist eine vertraute Kollegin, auf die sich alle Mitarbeiter verlassen können.*
value	schätzen
well-being *I'm their boss and I'm interested in their well-being.*	Wohlbefinden *Ich bin ihr Chef und sorge mich um ihr Wohlbefinden.*
wishy-washy *I'm sorry, this is a wishy-washy answer. Please be more precise.*	vage *Es tut mir leid, aber diese Antwort ist vage. Bitte sind Sie etwas präziser.*

12.5 Übungen – Exercise time

Task 1: Can you find the missing nouns?

1. We should keep t_of birthdays, births and anniversaries.
2. What sort of training would you b_from?
3. What are your strengths and w_for this task, do you think?
4. You should get to know your s_individually.
5. It could be a matter of the daily a_in your company.

Task 2: Which definition fits the term?

1. employee	(a) the entity that offers the job
	(b) the individual who takes on the job
2. brainstorming	(a) an individual solution to a question or issue
	(b) an unranked series of spontaneous ideas
3. monitor	(a) watch something or someone closely
	(b) record the details of an issue
4. biased	(a) open to ideas
	(b) closed for personal reasons
5. genuine	(a) it is general
	(b) it is real and honest

Task 3: Put in the correct form of the word in brackets.

1. They don't get it. I feel completely ________________. (understand)
2. We need to have a closer ________________ to our customers. (disconnect)
3. The company doesn't need to know that. It's ________________. (important)
4. I don't know why, but we have a problem to ________________ the staff. (demotivate)
5. You'll need quite a lot of ________________ working for this company. (inflexible)

Task 4: Some expressions used in the example speech for you to complete.

1. talk ________________ (fachsimpeln, über die Arbeit reden)
2. take ________________ (die Führung übernehmen)
3. work ________________ the ________________ of the team (im Rahmen des Teams arbeiten)
4. no ________________teamwork (kein Ersatz für Teamarbeit)
5. have a ________________ the company (Anteil nehmen an der Firma)
6. work for the ________________ good (für das allgemeine Wohl arbeiten)
7. efficiency ________________ a high ________________ (Leistung auf hohem Niveau)
8. ________________ the restaurant (das Restaurant vergrößern)
9. ________________ motivation (Motivation wecken)
10. ________________ the bonus scheme (das Prämiensystem reorganisieren)
11. ________________ co-operation (auf Kooperation angewiesen sein)
12. suggestions for ________________ efficiency (Vorschläge, um die Leistung zu steigern)

ARBEITSMITTEL AUF MYBOOK.HAUFE.DE

Auf der Internetseite zum Buch finden Sie weitere Übungen zu diesem Modul sowie eine Mustervorlage für ein Dankschreiben an Ihre Mitarbeiter:innen. Nachdem Sie die Vorlage angepasst haben, können Sie das Dankschreiben ausdrucken oder per E-Mail versenden.

13 Konflikte am Arbeitsplatz erkennen und lösen

Zusammenfassung
Konflikte am Arbeitsplatz sind leider keine Seltenheit. Konkurrenzkampf, Zeitdruck, aber auch einfach die fehlende Chemie zwischen Kollegen oder dem Vorgesetzten können Konflikte verursachen. In diesem Modul lernen Sie, wie Sie solche Situationen frühzeitig erkennen, damit umgehen und Konflikte lösen können. Denn nicht zuletzt werden ungelöste Konflikte zur Belastung, erzeugen Stress und Ängste, schaden dem Wohlbefinden. Dies hat letztlich auch negative Auswirkungen auf das Unternehmen.

Nützliche Gesprächsleitfäden und Checklisten helfen Ihnen dabei, Konflikte am Arbeitsplatz souverän zu meistern.

Die Meilensteine: Konfliktsituationen erkennen und bewältigen

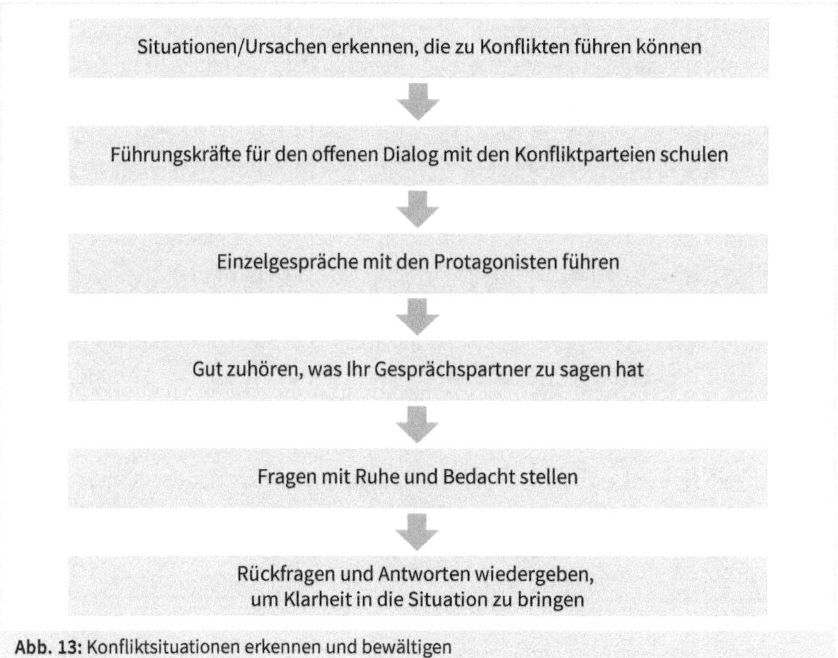

Abb. 13: Konfliktsituationen erkennen und bewältigen

13.1 Konfliktsituationen am Arbeitsplatz

Where there are people, there are conflicts. Unresolved conflicts can develop into crisis situations for leaders, teams and even the whole company. The working environment can become seriously poisonous if managers permit conflict to grow. They have to confront it head-on. But this can be tricky. Often a manager does not have the personal, deeper knowledge of the protagonists about the situation. He or she may not know if their solutions will cause further waves in the overall organisation.	Wo es Menschen gibt, gibt es auch Konflikte. Ungelöste Konflikte können zu einer Krise für Führungskräfte, Teams und sogar das gesamte Unternehmen eskalieren. Wenn Führungskräfte Konflikte zulassen, kann dies das Arbeitsumfeld erheblich vergiften. Sie müssen sich den Konflikten stellen. Dies kann jedoch recht schwierig sein. Oftmals fehlen der Führungskraft die persönlichen, tieferen Kenntnisse der Protagonisten über die Situation. Auch ist unklar ob seine/ihre Lösungen weitere Wellen innerhalb des gesamten Unternehmens schlagen werden.
When we work we may get stressed because of the quantity of work, the type of work, our lack of understanding about the work, the fact we must meet deadlines as well as a series of unexpected problems. Stress can make ill, physically and also mentally. However, perhaps the most stressful and stress-creating issues are not the work per se, but rather situations created by colleagues, bosses and customers.	Gründe, warum uns Arbeit stressen kann, sind das Arbeitsvolumen, die Art der Arbeit, unser fehlendes Verständnis für die Arbeit, die Tatsache, dass wir Termine einhalten müssen sowie eine ganze Reihe unerwarteter Probleme. Stress kann krankmachen, sowohl physisch als auch psychisch. Jedoch der stressigste und Stress verursachende Umstand ist nicht die Arbeit als solche, sondern vielmehr sind es Situationen, die von Kollegen, Vorgesetzten und Kunden verursacht werden.
These conflicts can be of many types – it could be a person who has a grievance, a problem or misunderstanding between a worker and a manager or issues between co-workers. The result is nearly always the same – conflicts are bad for business and certainly reduces the company's productivity.	Es gibt viele verschiedene Konstellationen für Konflikte – es kann sich um eine Person mit einer Beschwerde, mit einem Problem oder um ein Missverständnis zwischen Kollegen handeln. Das Resultat ist fast immer das gleiche – Konflikte sind schlecht für das Geschäft und führen unumstößlich zu Produktivitätseinbrüchen im Unternehmen.

13.1.1 Checkliste: Typical causes of conflict

Too much extra work.	Zu viel zusätzliche Arbeit.
An unresolved problem from an earlier conflict.	Ein ungelöstes Problem aus einem früheren Konflikt.
Some sort of harassment or bullying.	Belästigungen und Mobbing jeder Art.
The work environment is not up to standard.	Das Arbeitsumfeld entspricht nicht den Anforderungen.
The staff don't have equal opportunities for development or promotion.	Die Mitarbeiter:innen haben nicht dieselben Möglichkeiten zur Weiterentwicklung oder Beförderung.
Tasks are distributed without clear guidelines.	Aufgaben werden ohne klare Vorgaben vergeben.
There is unclear communication.	Es fehlt an klarer Kommunikation.
Somebody is treated unfairly.	Jemand wird ungerecht behandelt.
The management does not have sufficient or appropriate training.	Führungskräfte sind nicht ausreichend oder passend geschult.

13.1.2 Textbausteine: Talking about conflict in general

In diesem Leitfaden finden Sie eine Reihe von Musterformulierungen, um Konflikte anzusprechen und über sie zu reden.

Causes of stress	**Ursachen von Stress**
What's the cause of your stress?	Was ist die Ursache für Ihren Stress?
In your view what is putting you under pressure?	Was setzt Sie Ihrer Meinung nach unter Druck?
What disturbs you in this particular situation?	Was stört Sie in dieser spezifischen Situation?
Symptoms of stress	**Symptome von Stress**
What does it mean for you?	Was bedeutet das für Sie?
Does it mean you don't want to come to work?	Bedeutet es, dass Sie nicht zur Arbeit gehen möchten?
How does this stress affect you?	Wie wirkt sich Stress auf Sie aus?
I don't want to see or talk to anybody.	Ich möchte niemanden sehen und mit niemanden sprechen.

Cause of the conflict	Konfliktursache
Do you feel unfairly treated?	Fühlen Sie sich ungerecht behandelt?
I do a good job, but nobody sees it.	Ich mache einen guten Job, aber niemand sieht das.
My colleague gets all the good jobs.	Mein Kollege bekommt alle guten Jobs.
What have you got against Jim?	Was haben Sie gegen Jim?
They don't take me seriously because I'm a woman.	Die nehmen mich nicht ernst, weil ich eine Frau bin.
Talking about communication	**Über Kommunikation reden**
Somehow, we don't see eye to eye.	Irgendwie sind wir nicht einer Meinung.
He doesn't seem to know what he's talking about.	Er scheint nicht zu wissen, über was er da redet.
His instructions are so wishy-washy.	Seine Anweisungen sind so verwirrend.
She just interrupts and doesn't want to listen.	Sie unterbricht einfach und möchte nicht zuhören.
Equal opportunities	**Chancengleichheit**
It's not what you know, it's who you know.	Es geht nicht darum, was man weiß, sondern wen man kennt.
I'm sure I'm being paid less than him for the same work.	Ich bin mir sicher, dass ich für die gleiche Arbeit weniger verdiene als er.
The others get the chances. Nobody even asks me. It's frustrating!	Die anderen bekommen ihre Chance. Mich fragt man schon gar nicht. Es ist frustrierend!

13.2 Konfliktsituationen verhindern

Prevention is better than cure. The management with the help of Human Resources can minimise conflict by identifying and monitoring potentially threatening conflicts. Some tips for minimising these causes are as follows:

Vorbeugung ist die beste Medizin. In Zusammenarbeit mit dem Personalwesen können Führungskräfte Konflikte auf ein Minimum reduzieren, indem sie potenziell drohende Konflikte identifizieren und überwachen. Nachfolgend finden Sie einige Tipps, wie Sie solche Ursachen minimieren können:

Management can:
- train managers to handle potentially volatile discussions with their employees,
- support employees in expressing their opinions openly,
- emphathise with employees, accept the importance of employee's feelings,
- develop real listening skills,
- optimise a procedure to deal with grievances and disputes,
- also think about the possibility of an independent third party as a mediator for a difficult situation.

Das Management kann:
- Führungskräfte darin schulen, wie sie potenziell brisante Gesprächen mit Mitarbeiternden führen,
- Mitarbeiternde darin unterstützen, ihre Meinungen offen zu äußern,
- sich in die Lage von Mitarbeiternden versetzen und deren Gefühle von ernst nehmen,
- lernen wir man aktiv zuhört,
- die Vorgehensweise beim Umgang mit Beschwerden und Streitigkeiten optimieren,
- über die Möglichkeit nachdenken, bei schwierigen Situationen einen unabhängigen, externen Vermittler hinzuzuziehen.

13.2.1 Textbausteine: Konflikte schlichten

In der nachfolgenden Liste finden Sie zahlreiche Textbausteine, die Ihnen helfen, Konfliktgespräche diplomatisch zu führen.

Introducing the problem	**Das Problem ansprechen**
Colleagues have heard/mentioned that you are always arguing.	Kollegen haben gehört/erwähnt, dass Sie sich immer streiten.
It's come to my notice that you are not working well together.	Es ist mir zu Ohren gekommen, dass Sie nicht gut zusammenarbeiten.
We seem to be developing some issues among the team.	Wie es mir scheint, entstehen hier ein paar Probleme im Team.
Denying the problem	**Das Problem leugnen**
We weren't arguing at all.	Wir haben überhaupt nicht gestritten.
Oh, it was just a tiff, nothing serious.	Ach, es war nur eine kleine Meinungsverschiedenheit, nichts Ernsthaftes.
We just got a bit heated. It will blow over.	Wir sind nur etwas in Wallung geraten. Das beruhigt sich schon wieder.
There's no real problem. We're just a bit tired.	Es gibt nicht wirklich ein Problem. Wir sind nur ein wenig müde.

Attacking phrases	Attackierende Kommentare
What a cheek you've got!	Das ist eine Frechheit!
How dare you say that!	Wie können Sie es wagen, das zu sagen!
You should keep your opinion to yourself.	Sie sollten Ihre Meinung für sich behalten.
Well, mind your own business!	Kümmern Sie sich um Ihren eigenen Kram!
Defending yourself	**Sich verteidigen**
Well, I've worked here longer than you.	Also, ich arbeite hier schon länger als Sie.
I've got special qualifications in this field.	Ich verfüge über besondere Qualifikationen auf diesem Gebiet.
Perhaps my work's more valuable than other people's.	Vielleicht ist meine Arbeit ja wertvoller als die anderer Leute.
Defusing the situation	**Die Situation entschärfen**
If you could just listen a minute.	Wenn Sie mir eine Minute zuhören könnten.
Okay, you two, just calm down a bit!	Okay, Sie beide, Sie sollten sich etwas beruhigen.
So, it seems there is another problem after all?	So wie es aussieht, gibt es doch ein anderes Problem.
Please, get control of yourselves!	Bitte beherrschen Sie sich!
Well, we can certainly look into that.	Wir können dieser Sache sicherlich nachgehen.
Giving support as manager	**Unterstützung als Führungskraft geben**
Okay, I can give you both support with …	Okay, ich kann Sie beide bei … unterstützen.
Perhaps we can have a short talk behind closed doors.	Vielleicht können wir uns kurz hinter verschlossenen Türen unterhalten.
I'd certainly like to get to the bottom of it.	Ich möchte der Sache auf jeden Fall auf den Grund gehen.
Let's find a solution together.	Lassen Sie uns gemeinsam eine Lösung finden.

13.2.2 Beispieldialog: The feuding colleagues

Background: Mr Les Sharp, supervisor of the translation department of a language services company, has a discussion with Gertrud Holbein, a translator, and David Schwarz, also a translator.

Les Sharp: So, colleagues have noticed that you two have been arguing again. They quite rightly say it disturbs the atmosphere in our section. We have to nip this in the bud.

Gertrud Holbein: We weren't arguing. I was only informing David that sometimes his translation work isn't as accurate as it should be.

David Schwarz: That's a real cheek! How can you allow yourself to comment on my work? You're always poking your nose into my affairs. You're no more qualified than I am.

Gertrud Holbein: David, in case you haven't noticed I've worked here twice as long as you have.

David Schwarz: That has nothing to do with it. You always think your jobs and your work is more important than other people's.

Les Sharp: Okay, you two! Now just calm down a bit! Let's get control of ourselves and stop raising our voices. Think of your other colleagues!

Gertrud Holbein: Well, he can't get away with that.

Les Sharp: Just relax! Nobody wants to get away with anything.

Gertrud Holbein: I …

Les Sharp: Listen to me! It's my responsibility to get to the bottom of what's really going on here. So, both of you get hold of yourselves! Now, Gertrud, what do you find inaccurate with David's work.

Gertrud Holbein: Oh, I don't know. I feel I need to keep an eye on him.

David Schwarz: Control me, you mean.

Les Sharp: Calm down, David! So, it seems it's not a matter of inaccuracy or poor work?

David Schwarz: To tell the truth I feel Gertrud thinks she's better or more valuable than the rest of us.

Les Sharp: What makes you say that? Can you give an example?

David Schwarz: It seems to us she gets the more interesting jobs.

Les Sharp: Well, that's something we can certainly look into.

David Schwarz: And she's treated differently than other members of staff. She's a personal friend of some members of the management.

Les Sharp: So that seems to be a different issue. What's your comment on that one, Gertrud?

Gertrud Holbein: Well, I hadn't thought about that before. It hadn't crossed my mind.

Les Sharp: Okay. So, this is something I can give you both support with. And I'll have an unofficial talk with those people involved. Perhaps I could have a short talk with each of you individually?

13.3 Konflikte lösen

The goal of resolving a conflict is not to decide which person is right or wrong but rather to reach a solution that everyone can live with. To manage conflict effectively you need to create an open communication environment. Listening is key here. Make sure that you understand what the issue is by asking questions. Before any kind of problem-solving can take place it is important that you allow the protagonists to express their feelings and you should acknowledge them.	Beim Lösen von Konflikten geht es nicht darum zu entscheiden, wer Recht hat und wer nicht, sondern vielmehr darum, eine Lösung zu finden, mit der alle leben können. Um mit Konflikten effektiv umzugehen, müssen Sie ein offenes Kommunikationsumfeld schaffen. Die Basis hierfür ist das Zuhören. Stellen Sie sicher, dass Sie wissen, was das Problem ist, indem Sie Fragen stellen. Bevor überhaupt mit dem Lösen von Problemen begonnen werden kann, ist es wichtig, dass Sie den Protagonisten die Möglichkeit geben, ihre Gefühle zum Ausdruck zu bringen and zu akzeptieren.
Establish the reason for the conflict and the impact on work or relationships. For this it might be necessary that you talk to employees separately and question them about the situation. Once you have established the reason of conflict you should try to find common areas of agreement, such as agreeing on actions to be taken or agreeing on small changes in procedure.	Ermitteln Sie die Ursache für den Konflikt und die Auswirkungen auf die Arbeit und die Beziehungen. Hierfür kann es erforderlich sein, dass Sie mit Mitarbeitern:innen Einzelgespräche führen und sie über die Situation befragen. Sobald die Konfliktursache feststeht, sollten Sie versuchen, gemeinsame Bereiche der Übereinstimmung zu finden, und sich zum Beispiel darauf einigen, welche Maßnahmen ergriffen werden sollen oder welche Änderungen in der Vorgehensweise umgesetzt werden sollen.

You may want to fix a follow-up meeting to see what the conflict parties are doing. If you feel that the conflict has still not been resolved, you may need to bring in an outside facilitator.

Eventuell sollten Sie ein Folgemeeting vereinbaren, um zu sehen, wie es zwischen den Konfliktparteien läuft. Wenn Sie das Gefühl haben, dass der Konflikt nach wie vor besteht, kann das Einschalten eines externen Schlichters erforderlich sein.

13.3.1 Checkliste: Tipps für den Umgang mit Konfliktsituationen

Die folgenden Tipps bieten gute Hilfestellungen im Umgang mit Konflikten.

- Listen to both sides with understanding and empathy!
- Encourage the 2 parties to work it out, with a little help from yourself.
- Nip it in the bud before it spreads through the team.
- Try to identify the real problem!
- Check the rules for procedures yourself!
- Find a positive solution
- The whole discussion and solution should be documented.
- Help them to communicate more clearly and fairly with you and each other!

A	**Give the people the chance to tell their story**
When members of staff are angry and nervous introduce the idea of mediation so that they slow down and see the sense of it. They must explain the depth of the conflict to their manager.	
B	**Bring the protagonists down to earth**
Explain the bigger picture to them. They are probably too fixed on their own issues.	
C	**Get down to the real point**
Identify what the real reason for the protagonists' issue is.	
D	**Be impartial**
Don't be influenced by the politics in the workplace. When you intervene be seen to take no side. Be fair and impartial! If you cannot understand the quarrel and its causes, you will only add oil to the flames. Keep all sides focussed on solving the problem!	

Typical myths about workplace conflict

- **Conflict is always bad so we must avoid it.**
  This isn't true. The worst thing is not to discuss. It will then grow and grow. Better to confront it but sensitively.

- **Difficult employees are the reason for conflict.**
  More often it's because the expectations of colleagues or bosses are not realistic. It can lead to misunderstanding and chaos.
- **The reason for the conflict is obvious.**
  Rarely! First we have to identify the real or deep issues behind it. We need dialogue and communication.
- **It's always the manager's job to clarify the problem.**
  He or she should be the guiding, impartial light. But the protagonists need to master their own conflict management for future situations.

13.4 Der Manager in Konfliktsituationen

What sort of a manager are you? Good question! Are you skilful in communicating fairly with the protagonist in a conflict situation? Our image of ourselves is often very different from how we are seen by others. In a stressful confrontation some managers react by using sarcasm against employees. Perhaps they believe they are being clever and funny with quick witty answers but the employees who are being spoken to may feel foolish and naive.

Was für eine Art von Führungskraft sind Sie? Gute Frage! Sind Sie geschickt darin, in einer Konfliktsituation ein faires Gespräch mit einem Protagonisten zu führen? Das Bild, das wir von uns selbst haben, unterscheidet sich oftmals stark von dem Bild, das andere von uns haben. In einer stressigen Konfrontation reagieren manche Führungskräfte mit Sarkasmus gegenüber Mitarbeitenden. Vielleicht glauben sie, dass sie mit ihren schnellen und witzigen Antworten clever und komisch sind, aber die betroffenen Mitarbeiter:innen können sich dumm und naiv vorkommen.

Another approach is also based on human nature. A manager might prefer to not confront employees who have a certain strength of character and to avoid trouble with this person. Instead he might prefer to shout orders at or pick on an employee who is less dominant, more submissive. But beware! Other colleagues may notice this and label that manager a bully and a coward.

Ein weiterer Ansatz beruht ebenso auf der menschlichen Natur. Eine Führungskraft zieht es eventuell vor, charakterstarken Mitarbeitenden aus dem Weg zu gehen und eine Konfrontation zu vermeiden. Stattdessen sucht er:sie sich lieber Mitarbeiter:innen die weniger dominant und stärker autoritätsgläubig sind, um sie herumzukommandieren oder auf ihnen herumzuhacken. Aber Vorsicht! Anderen Kollegen:innen kann das auffallen und sie stempeln diese Führungskraft als Tyrann:in und Feigling ab.

Some managers have an attitude which includes »Oh, I've been promoted to this high level. Well, I'm the one who knows best in all situations and about all things.« If this comes out in the manager's attempt to resolve a conflict, most probably all involved employees will become silent and avoid developing any creative or useful ideas with this manager. They will certainly criticise him or her privately amongst themselves.
The bottom line is that the key to managing conflict is listening, fair and sensitive comment and a certain empathy - a viewing of the world as »we« rather than »I«

Manche Führungskräfte haben die Einstellung: »Nun, ich wurde auf diese hohe Position befördert. Dann bin ich auch derjenige, der in allen Situationen und in allen Fragen am besten Bescheid weiß.« Sollte sich diese Haltung zeigen, wenn es darum geht, Konflikte zu lösen, dann werden höchstwahrscheinlich alle Mitarbeiter:innen verstummen und es vermeiden, kreative oder nützliche Ideen mit dieser Führungskraft zu entwickeln.
Die Voraussetzung zur Bewältigung von Konflikten sind folglich Zuhören, faire und einfühlsame Kommentare sowie eine gewisse Empathie - eine Wir- statt Ich-Sichtweise.

13.4.1 Situationen deuten

Let's imagine you're the manager in a department. There is stress, we have a workplace conflict. Can you interpret the situation to begin to reach the real problem?

Reaction of employee	...
He or she's not speaking.	You are informing the worker what you have heard or seen. Your information is probably authoritative. The employee is silent. He says nothing. This does not indicate that he does not accept, that he resists. He may be considering the news. You must carefully form your questions and information to reassure him or her. You are there to help, not to punish.
He or she's in tears.	You told your employee that his work wasn't good enough. You explain that he wasn't up to it. But suddenly he bursts into tears. Clearly / he was shocked by this news. How do you react? Pity him? No, this is counter-productive. Explain to him that it's not about his job. That you should meet to find a good way forward to correct things.
He or she laughs about it.	You make a criticism. The employee laughs. But perhaps he is not making a joke of things. It is not that he is not taking the criticism seriously. Maybe it's a matter of nerves. In any case you should not laugh along with him. Be serious and firm! This will help him to stop laughing and listen.
Yes, it's my fault.	The employee keeps admitting his guilt. He doesn't look you in the eye. He's probably hoping his constant apologising will help him escape from the situation. You have to focus on the real situation, be fair and firm and get him to accept your position and commit.

13.5 Vokabeln – Vocabulary input

English	German
add oil to the flames	Öl ins Feuer gießen
admit guilt	Schuld eingestehen
approach	Ansatz
appropriate	passend, geeignet
attitude	Einstellung
authoritative	bestimmend
bottom line; the ~ is	im Endeffekt, unterm Strich
bring somebody down to earth	jemand auf den Boden der Tatsachen zurück-holen
bully somebody	jmdn. schikanieren, einschüchtern, mobben
bullying	Schikane, Mobbing
but beware	doch Vorsicht; aber aufgepasst
calm down	sich beruhigen
cause waves	Wellen schlagen
clarify	aufklären
commit to something *We commit to strengthening our cooperation with our partners.*	sich zu etwas verpflichten *Wir verpflichten uns, die Zusammenarbeit mit unseren Partnern zu stärken.*
conflict	Konflikt
counter-productive	kontraproduktiv
coward	Feigling
co-worker	Kollege
cross one's mind	jmdn. in den Sinn kommen
deadline; meet the ~	Frist einhalten
deny *The employee denied any wrongdoing.*	bestreiten, leugnen *Die Mitarbeiterin hat jegliches Fehlverhalten bestritten.*
denial	Ableugnen, Bestreiten
disturb	stören
disturbance	Störung

English	German
document, to ~ something	etw. dokumentieren
documentation	Dokumentation
empathy	Empathie; Einfühlungsvermögen
empathise with somebody	sich in jmdn. einfühlen
encourage *My boss encouraged me to apply for the vacancy.*	ermutigen *Mein Chef hat mich ermutigt, mich für die freie Stelle zu bewerben.*
environment	Umfeld; Umwelt
equal opportunities	Chancengleichheit
express an opinion	seine Meinung kundtun
feud	Streit; Fehde
foolish	dumm
get away with …	davonkommen
get control of oneself	sich beherrschen
get to the bottom of *We must get to the bottom of what causes this high employee turnover rate.*	Dingen auf den Grund gehen *Wir müssen der Sache auf den Grund gehen und herausfinden, was diese hohe Mitarbeiterfluktuationsrate verursacht.*
grievance	Beschwerde; Missstand
harass	belästigen
harassment	Belästigung
head-on	sich etwas stellen
heated	in Wallung geraten
image of oneself / self-image	Selbstverständnis
impartial	unparteiisch; objektiv
inaccurate	ungenau
inaccuracy *We are not liable for any inaccuracy relating to information contained in this document.*	Ungenauigkeit *Wir übernehmen keine Haftung für jegliche Ungenauigkeiten hinsichtlich der in diesem Dokument enthaltenen Informationen.*
indicate	auf etwas hindeuten
intervene	eingreifen
intervention	Intervention; Eingreifen

English	German
issue *We must take the issue of unequal treatment seriously.*	Angelegenheit; Problem *Wir müssen das Problem der ungleichen Behandlung ernst nehmen.*
laugh about something	über etwas lachen
mediate	schlichten; vermitteln
mediation	Schlichtung; Vermittlung
mediator	Schlichter; Vermittler
Mind your own business!	Kümmern Sie sich um Ihren eigenen Kram!
minimise	minimieren
myth	Gerücht; Mythos
obvious	offensichtlich
offender	Angreifer; Täter
open dialogue	offener Dialog
permit	erlauben
pick on somebody *My boss is always picking on me.*	auf jmdn. herumhacken *Mein Chef hackt immer auf mir herum.*
poke one's nose into something	sich in etwas einmischen
productivity	Produktivität
promote	befördern
promotion	Beförderung
protagonist	Protagonist; Akteur
punish	bestrafen
punishment	Bestrafung
quarrel	Streit; Disput
rarely	selten
rephrase something	etwas umformulieren
sarcasm	Sarkasmus
see eye to eye *The are working well together even though they don't always see eye to eye.*	einer Meinung sein *Sie arbeiten gut zusammen, auch wenn sie nicht immer einer Meinung sind.*
skilful	geschickt

English	German
stressful	stressig
stress-creating	Stress verursachend
submissive	autoritätsgläubig; unterwürfig
supervise	beaufsichtigen; überwachen
supervisor	Vorgesetzter:e
threaten	drohen
take sides	Partei ergreifen
treat unfairly	unfair behandeln
tricky *Personnel managers face the tricky task of filling a vacancy with suitable employees.*	schwierig; heikel *Personalmanager haben die schwierige Aufgabe, freie Stellen mit geeigneten Mitarbeitern zu besetzen.*
trigger; to ~ something	etwas auslösen
valuable	wertvoll
volatile	brisant; unberechenbar
volatility	Unbeständigkeit
What a cheek!	Was für eine Frechheit!
work environment *Performance will be negatively affected in a hostile work environment.*	Arbeitsumfeld *Ein von Anfeindungen geprägtes Arbeitsumfeld wirkt sich negativ auf die Leistung aus.*

13.6 Übungen – Exercise time

Task 1: What are the nouns from the following words?

1. produce = ________________
2. harass = ________________
3. clear = ________________
4. disturb = ________________
5. interrupt = ________________
6. prevent = ________________
7. 7. encourage = ________________

Task 2: Put the words in brackets into the sentences.

1. The whole discussion should be ________________. (document)
2. The manager must try to get a deeper ________________ of the situation. (know)
3. He has to be ________________ towards the protagonists. (impartiality)
4. Give only fair and ________________ comments to employees. (sensitive)
5. How do you ________________ to difficult employees? (reaction)

Task 3: Find terms for the following definitions.

1. an argument between people in the company → w________________ c________________
2. say sorry to somebody → a________________
3. a negative comment about somebody → c________________
4. jobs are given to various workers → d________________
5. a co-worker more or less at the same level → c________________

14 Entsendung ins Ausland

Zusammenfassung

Eine Entsendung ins Ausland kann für die Karriere eines Mitarbeiters ein großes Plus bedeuten. Für die Personalabteilung gibt es jedoch dazu in der Vor- und Nachbereitung zahlreiche Dinge zu regeln. Zu den wichtigsten Punkten geben wir Ihnen in diesem Modul Tipps und Arbeitsmittel an die Hand, damit Sie alles sicher im Griff haben.

- *Checkliste* für die Auswahl des für die Entsendung optimalen Mitarbeiters.
- *Gesprächsleitfaden* für die Begrüßung eines entsendeten Mitarbeiters.
- *Checkliste* und *Gesprächsleitfaden* für die Reintegration eines entsendeten Mitarbeiters.

Am Ende des Moduls finden Sie alle wichtigen *Vokabeln* zu diesem Thema und *Übungsaufgaben*, mit denen Sie Ihre Kenntnisse überprüfen können.

Darüber hinaus finden Sie im Anhang des Buches hilfreiche Informationen zur Kommunikation in fremden Ländern.

Die Meilensteine: Von der Entsendung bis zur Reintegration

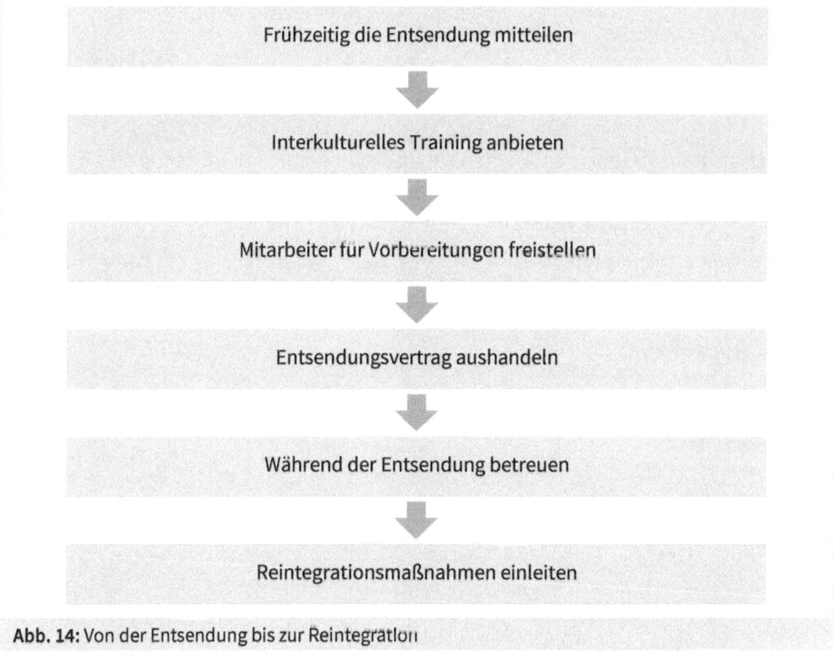

Abb. 14: Von der Entsendung bis zur Reintegration

14.1 Gründe für Entsendungen und Kriterien für die Mitarbeiterauswahl

There are several reasons why companies should send selected employees to foreign subsidiaries for a longer period of time. Below you will find a list of such reasons:

- to establish a new venture
- to stimulate an international team of managers
- to fill a vacancy because local staff don't have the qualifications
- to train local employees
- to have influence over the local operation
- to offer a development chance to a high potential

Es gibt verschiedene Gründe, warum Firmen ausgewählte Mitarbeiter:innen für einen längeren Zeitraum zu ausländischen Niederlassungen schicken sollten. Nachfolgend finden Sie eine Auflistung solcher Gründe:

- um eine neue Firma zu gründen
- um ein internationales Team von Managern zu fördern
- um vor Ort eine aufgrund mangelnder Qualifikation von Mitarbeitenden vor Ort offene Position zu besetzen
- um Mitarbeitende vor Ort weiterzubilden
- um Einfluss auf das operative Geschäft vor Ort zu haben
- um einem Potenzialträger die Möglichkeit der Weiterentwicklung zu bieten

14.1.1 Checkliste: Mitarbeiter:innen für Entsendungen auswählen

What is the employer looking for when selecting an employee for secondment? There seem to be five areas which are important to decision-makers.

1. Job know-how and motivation		
High motivation		
Energy and initiative		
Organizational ability		
Managerial ability		
2. Personal skills		
Politeness and tact		
Empathy		
Respect		

3. Adaptability/flexibility		
Adaptable		
Ability to deal with stress		
Willing to change		
4. Openness to cultural matters		
Interest in foreign cultures		
Outgoing and extrovert		
Openness		
5. Family situation		
How adaptable is the wife/husband?		
Wife's/husband's positive feeling		
Stable marriage		

14.1.2 Gesprächsleitfaden: Ein beratendes Gespräch führen

Dieser Leitfaden enthält Musterformulieren, wie Sie jemandem Ratschläge geben können. In dem anschließenden Beispieldialog geht es um das Thema Personalentsendung ins Ausland.

Introducing a topic in general	**Allgemeine Einleitung in ein Thema**
As you read in my letter we're thinking of sending you to …	Wie Sie in meinem Schreiben gelesen haben, dachten wir, Sie nach … zu entsenden.
In general we'd like/prefer to …	Im Allgemeinen würden wir gerne …
To give a few rough guidelines we …	Um ein paar grobe Vorgaben zu nennen, wir …
We need a few broad tips on how to …	Wir brauchen einige allgemeine Hinweise wie …
Offering to give information	**Informationen anbieten**
Well, there are a few things to keep in mind. Shall I give you a short overview?	Nun, da gibt es ein paar Dinge zu beachten. Soll ich Ihnen einen kurzen Überblick geben?
I can show you what I mean, if you like.	Wenn Sie möchten, kann ich Ihnen zeigen, was ich meine.
If you're okay with it I could just give you a quick summary.	Wenn Sie damit einverstanden sind, könnte ich Ihnen eine kurze Zusammenfassung geben.
I could show you the basics.	Ich könnte Ihnen das Grundlegende zeigen.

Listing several items	**Verschiedene Punkte anführen**
Well, the first is the …, and then the …	Nun, zuerst ist die … dann die …
First of all we should think about … Later we can …	Zuallererst sollten wir über … nachdenken. Später können wir …
There are several points. Firstly. you'll get … After … months you'll have … and finally there'll be …	Da gibt es verschiedene Punkte. Zuerst werden Sie … bekommen. Nach … Monaten haben Sie … und zuletzt wird es … geben.
Giving more detailed advice	**Eingehende Ratschläge geben**
So, to go into detail, you'll need …	So, um auf die Einzelheiten einzugehen, Sie werden …
That was it in general, but we should focus on …	Das war es im Allgemeinen, aber wir sollten uns auf … konzentrieren.
Well, I'd like to add a few things here to clarify.	Nun, ich würde gern einige Dinge zur Klärung hinzufügen.
Giving final advice	**Abschließenden Ratschlag geben**
Well, this is the last major point.	Nun, das ist der letzte Hauptpunkt.
Finally I'd like to warn you that …	Zuletzt würde ich Sie gerne vor … warnen.
So, at the end of the day this is my advice to you.	Nun, letzten Endes ist dies mein Rat an Sie.
To my knowledge, this would be the best in the end.	Nach meinem Wissen wäre das am Ende das Beste.

14.1.3 Beispieldialog: Seconding an employee abroad

Background: Discussion between Dan Loan, the recruiter from the personnel department, and the legal HR consultant Anne Horovitz from the USA.

Dan Loan: So, as you read in my letter we're thinking of sending one of our employees abroad. I guess there are a whole lot of legal issues to consider on this one?

Anne Horovitz: Of course, there are several things to keep in mind. Shall I give you a short overview?

Dan Loan: If you could.

Anne Horovitz: Well, the first item is the employee's full agreement. You should obtain his consent. No coercion! Then there are the visa and any other immigration requirements. You'll have all sorts of issues to manage concerning tax, where it should be paid etc, and there's the matter of social security.

Dan Loan: Already complicated enough. Any other major considerations?

Anne Horovitz: It's certainly important to consider the duration of the secondment. The terms and conditions that are in force. Will the employee have the same benefits as at home?

Dan Loan: Okay. My next question is »who has responsibility for the employee?«

Anne Horovitz: You have to negotiate and put this in writing in the secondment agreement. It's usual that the employee stays in the employment of the home employer when he or she is seconded to work for the company abroad. But his or her tasks will be managed by the overseas company. The right to dismiss or discipline stays with the home employer.

Dan Loan: What about salary and grievances?

Anne Horovitz: The home employer deals with salary and grievances. Holiday questions are normally dealt with by the overseas company.

Dan Loan: There could be a question of loyalty.

Anne Horovitz: Normally the loyalty is to the home employer. Any questions of intellectual property should be discussed and integrated in the agreement.

Dan Loan: There is sometimes a big problem when the employee returns to the home employer. What should happen after termination of the secondment?

Anne Horovitz: Okay. Well, this is the major point. The employee returns to the home country to resume work with the employer. If he or she's been away three or four years there may be no position for him or her. He or she may be made redundant.

It should be clearly written into the contract. You, the employer, should perhaps offer a specific position for the employee to return to.

14.2 Entsendete Mitarbeiter:innen begrüßen

Secondment is often linked to climbing the career ladder. Good managers try to build career bridges in order to open up new areas for experience and allow the manager to build on his or her skills.

Eine Entsendung ist oft mit einem Aufstieg auf der Karriereleiter verbunden. Gute Führungskräfte versuchen, Karrierebrücken aufzubauen, mit dem Ziel, sich neue Erfahrungsbereiche zu eröffnen und die eigenen Fähigkeiten weiter auszubauen.

With secondment it's a bit more complicated. There are various interpretations of it but generally employee leave their »main« job for a period of a couple of years to acquire new know-how and international experience. The secondment may be to another subsidiary or an overseas posting.

Mit der Entsendung ist es ein wenig komplizierter. Es gibt eine Vielzahl von Interpretationen, aber im Allgemeinen verlassen Mitarbeiter ihren Hauptarbeitsplatz für einige Jahre, um sich neues Wissen und internationale Erfahrungen anzueignen. Die Entsendung kann zu einer anderen Niederlassung oder Auslandsvertretung erfolgen.

It might appear to be a side-step or even a step down but should be seen as part of a larger strategy, so that after returning the employee can continue with the next stage of climbing the ladder. For this reason it's vital to keep strong links with the employer back home.

Dies mag den Anschein haben, dass man sich auf der Stelle bewegt oder sogar einen Schritt zurück macht, sollte aber als Teil einer größer angelegten Strategie gesehen werden, so dass Mitarbeiter:innen nach ihrer Rückkehr mit der nächsten Stufe der Karriereleiter fortfahren kann. Dafür ist es unerlässlich, gute Verbindungen mit dem Arbeitgeber zu Hause zu wahren.

14.2.1 Gesprächsleitfaden: Entsendete Mitarbeiter:innen begrüßen

Musterformulierungen für die Begrüßung eines:er neuen Mitarbeiters:in aus dem Ausland erhalten Sie in diesem Leitfaden. Direkt im Anschluss daran finden Sie ein einen Beispieldialog, in dem es um die Begrüßung eines:er entsendenten Mitarbeiters:in geht.

Welcoming to the company	In der Firma willkommen heißen
Well, Mike, I'm very pleased to welcome you to …	Nun, Mike, ich möchte dich bei … herzlich willkommen heißen.
It's my pleasure to welcome you here.	Es ist mein Vergnügen, Sie hier willkommen zu heißen.

I hope you'll feel at home here at …	Ich hoffe, Sie werden sich in … zuhause fühlen.
I'd like to give you a warm welcome to …	Ich möchte Sie bei … herzlich willkommen heißen.
We're very happy to have you here at … at last.	Wir sind sehr froh, Sie hier bei … zu haben.
Explaining the reason for the meeting	**Die Gründe für das Meeting erläutern**
Actually, that's part of why we're having this meeting.	Eigentlich ist dies ein Grund für dieses Meeting.
That's one reason why we've come together today.	Das ist ein Grund, warum wir heute zusammengekommen sind.
That's why we've asked for this appointment.	Deshalb haben wir Sie um diese Verabredung gebeten.
That's exactly what we wanted to discuss today.	Das ist genau ein Punkt, den wir heute besprechen wollten.
Making a comparison	**Vergleiche anstellen**
Well, at … I get most of my salary in my hand.	Nun, bei … bekomme ich den größten Teil meines Gehaltes bar ausbezahlt.
Well, I was certainly earning more than now.	Nun, ich habe sicherlich mehr verdient als jetzt.
My take-home pay was more than today.	Mein Nettoverdienst war höher als heute.
I was doing better than I am now.	Mir ging es besser als jetzt.
Explaining difficult matters	**Schwierige Punkte erläutern**
I know. The deductions are really high, but …	Ich weiß, die Abzüge sind wirklich hoch, aber …
I understand what you're saying, but have you ever thought …?	Ich verstehe, was Sie sagen, aber haben Sie jemals bedacht, dass …
That's right. Nevertheless we should perhaps think about …	Das ist richtig. Trotzdem sollten wir vielleicht über … nachdenken.
Yes, that's quite true. On the other hand we could …	Ja, das ist richtig. Andererseits könnten wir …

14.2.2 Beispieldialog: Discussion between Human Resources and new arrival

Background: Mike Novick, a senior engineer from the US has been transferred from the subsidiary in Chicago to the company headquarters in Germany. He is having a conversation with the department leader of HR, Bernd Schulz.

Bernd Schulz: Well Mike, I'm very pleased to welcome you to Heidelberg and to the headquarters. You were working at the subsidiary in Chicago, weren't you?

Mike Novick: That's right. And now I'm on tour for three years in Germany.

Bernd Schulz: And I hope you'll enjoy it. Your advantage is that you're on your own. I mean you don't have family with you.

Mike Novick: Family might be nice, but I heard family dissatisfaction is the biggest reason for the failure of these overseas postings.

Bernd Schulz: Yes, that's right. Actually, that's part of why we are having our meeting today. You've been here for three weeks already. How are you settling in? What are your impressions?

Mike Novick: Well, it's a nice little town. But the big picture is very different from back home.

Bernd Schulz: What do you mean exactly?

Mike Novick: Well, in Chicago I got most of my salary in my hand. Here it's more or less only the half.

Bernd Schulz: I know. The deductions are really high. But it pays for the social network.

Mike Novick: There we have it again. Back home everybody has to be responsible for him or herself.

Bernd Schulz: But here the state has its say in everything. You can't fight it. You just make problems for yourself.

Mike Novick: I realise that.

Bernd Schulz: Any other observations?

Mike Novick: A very nice thing for me is the number of days of vacation. In Chicago, I had two weeks. Here I've got thirty days. That's great! I'm going to do a whole lot of travelling.

Bernd Schulz: Well, yes. That won't be a problem.

14.3 Rückkehr nach der Entsendung

Returning home can be a shock after a seconded employee has lived in another culture for several years. Below is a checklist of suggestions for cushioning the blow.

Für entsandte Mitarbeiter:innen, die für einige Jahre in einer anderen Kultur gelebt hat, kann die Rückkehr nach Hause ein Schock sein. Es folgt eine Checkliste mit Vorschlägen, die zeigen, wie dieser Schock abgefedert werden kann.

14.3.1 Checkliste: Rückkehr nach der Entsendung vorbereiten

Checklist for re-entry	Checkliste für die Rückkehr
Take some time for switching the work and the social situations. Perhaps you could have a short holiday back in the place you were seconded to.	Nehmen Sie sich Zeit, sich auf den Wechsel der Arbeit und der sozialen Umstände vorzubereiten. Vielleicht können Sie einen kurzen Urlaub am Entsendungsort verbringen.
Think about the changes in the last three years. And try to imagine that things have also changed back home. Maybe fashion, politics and other things are not the same anymore.	Denken Sie an die Veränderungen in den letzten drei Jahren. Und versuchen Sie sich vorzustellen, wie sich die Dinge in Ihrem Heimatland verändert haben. Vielleicht sind Mode, Politik und andere Dinge nicht mehr dieselben.
Connect in a relaxed way with old friends and professional contacts. They may have new contacts and could be bowled over by you running in.	Stellen Sie auf eine entspannte Art und Weise Kontakt mit alten Freunden und Kollegen her. Sie haben vielleicht neue Kontakte und könnten überrascht sein, Sie zu treffen.
You should make a long-term and a short-term plan. For example, maybe qualifications, or setting up home again, or even changing career path.	Sie sollten einen langfristigen und einen kurzfristigen Plan machen. Beispielsweise hinsichtlich der Qualifikationen, wieder eine Bleibe zu finden oder sogar sich beruflich zu verändern.
You'll probably feel a culture shock in the areas of finance, organisational methods and the psychology in general. Maybe you'll have a salary reduction, maybe there's been a change in the corporate culture. Be prepared! Don't be shocked!	Was Finanzen, organisatorische Praktiken oder auch die Psyche im Allgemeinen anbelangt, werden Sie sicherlich einen Kulturschock erleben. Vielleicht verdienen Sie weniger, vielleicht gab es eine Änderung in der Firmenkultur. Seien Sie vorbereitet! Seien Sie nicht schockiert!

14.3.2 Gesprächsleitfaden: Auf die anstehenden Veränderungen vorbereiten

Sie führen ein Gespräch mit einem Mitarbeiter, der nach einem Auslandsaufenthalt an seinen Arbeitsplatz zurückkehren wird. Wie Sie das Gespräch aufbauen und was Sie sagen können, erfahren Sie hier.

Addressing the employee's time in the country	Über die Zeit von Mitarbeiter:innen im Land sprechen
So, you've been working here in Germany for almost three years.	So, Sie arbeiten hier in Deutschland seit fast drei Jahren.
I guess you've been here for about three years.	Ich glaube, Sie sind seit etwa drei Jahren hier.
Well, XX years is a long time.	Nun, XX Jahre sind eine lange Zeit.
So, how long have you been with us, about XX years, I guess?	Wie lange sind Sie bei uns, etwa XX Jahre, richtig?
Asking about concerns	**Nach Bedenken fragen**
Have you any worries about your re-entry?	Haben Sie hinsichtlich Ihres Wiedereinstiegs irgendwelche Bedenken?
Is there anything on your mind concerning ...?	Haben Sie hinsichtlich ... etwas auf dem Herzen?
Did you want to ask me anything about ...?	Möchten Sie mich etwas über ... fragen?
Expressing concerns	**Bedenken zum Ausdruck bringen**
Nevertheless, it's still worrying that ...	Nichtsdestotrotz ist es immer noch beunruhigend, dass ...
I know, but I can't help thinking that ...	Ich weiß, aber mir drängt sich der Gedanke auf, dass ...
I can't help feeling that ...	Ich kann mich des Eindrucks nicht erwehren, dass ...
There's something I need to get off my chest.	Da gibt es etwas, was ich mir von der Seele reden muss.
Excusing lack of action	**Tatenlosigkeit entschuldigen**
Yes, I know. That was out of our hands.	Ja, ich weiß. Hierauf hatten wir keinen Einfluss.
I'm very sorry, but we didn't have the power to ..., but ...	Es tut mir sehr leid, aber wir hatten nicht die Macht, es zu ..., aber ...
You're totally right, but our hands were tied.	Sie haben völlig Recht, aber uns waren die Hände gebunden.
I regret it very much, but we had no time at all.	Ich bedauere das, aber wir hatten überhaupt keine Zeit.

14.3.3 Beispieldialog: Discussion with an employee returning home

Background: Conversation between the HR manager Tom Lee and the employee Bill Lorey about the return to the employee's home base.

Tom Lee: So you've been working here in Germany for almost three years. I expect you're looking forward to getting back home.

Bill Lorey: Well, the work here was really interesting for me as a documentation manager. And being the only native speaker of English and trained copywriter made me feel a bit like a big fish in a small pond. I'm afraid that'll be different when I get back.

Tom Lee: I can imagine. Have you any worries about your re-entry?

Bill Lorey: Well, of course now I have to face the fact that I may not get the job I expect when I arrive.

Tom Lee: That's often the case. But I saw in your contract that HQ is obliged to give you an employment contract. That can still be a stepping stone.

Bill Lorey: Nevertheless it's still worrying. But my wife and kids will be glad to be home.

Tom Lee: I believe the company did its best to integrate them all. Your wife had two German courses, I believe?

Bill Lorey: True enough, but it takes more than that to get middle level competence in a language. And, of course, she had no chance to practise. And more importantly, no chance to work – no work permit.

Tom Lee: Yes, I know. That was out of our hands. But on the positive side, you have a much broader outlook than those colleagues who have never travelled. I know three years is a long time. If I'm not wrong, HR in Chicago has a sort of reintegration scheme for expatriate staff, doesn't it? In any case I've put together a checklist to give you a basis for your re-entry.

14.4 Vokabeln – Vocabulary input

Nachfolgend finden Sie eine Auflistung wichtiger Fachbegriffe rund um das Thema Mitarbeiterentsendung.

English	German
acquire	erlangen
adaptable *A good team must be adaptable.*	anpassungsfähig *Ein gutes Team muss anpassungsfähig sein.*
agreement	Vereinbarung
beware of *Beware of smiling bank mangers.*	sich in Acht nehmen; Vorsicht! *Nehmen Sie sich vor lächelnden Bankmanagern in Acht.*
coercion *There must be no coercion regarding the secondment.*	Zwang *Es darf keinen Zwang bezüglich der Entsendung bestehen.*
concern	Bedenken
consent *You need the consent of the supervisory board.*	Zustimmung *Sie brauchen die Zustimmung des Aufsichtsrats.*
contagious	ansteckend
cushion	abfedern
deduction	Abzug
dissatisfaction *The dissatisfaction amongst employees has risen.*	Unzufriedenheit *Die Unzufriedenheit in der Belegschaft ist gestiegen.*
distance learning	Fernstudium
empathy *She has shown tremendous empathy.*	Einfühlungsvermögen *Sie hat großes Einfühlungsvermögen gezeigt.*
expatriate staff	Mitarbeiter, die ins Ausland entsendet wurden
grievance	Beschwerde
hostility	Feindschaft

English	German
impression *First impressions count.*	Eindruck *Der erste Eindruck zählt.*
intellectual property	geistiges Eigentum
labels; attach ~ to	in eine Schublade stecken
loyalty	Loyalität
mistrust *I have a certain degree of mistrust for banks.*	Misstrauen *Ich hege ein gewisses Misstrauen gegenüber Banken.*
observation	Beobachtung
openness	Offenheit
outgoing	aufgeschlossen
pleasure	Vergnügen
posting *The period of the posting must not exceed 5 years.*	Versetzung, Entsendung *Die Dauer der Entsendung darf 5 Jahre nicht übersteigen.*
prejudice	Vorurteil
reinforce	bestärken
settle in	sich einleben
side-step	Schritt zur Seite
social security	Sozialversicherung
step down *This could be a step down in his career.*	Schritt zurück *Dies könnte ein Schritt zurück in seiner beruflichen Laufbahn sein.*
stepping stone	Sprungbrett
subsidiary	Tochtergesellschaft
take-home pay	Nettoverdienst
terms and conditions	Geschäftsbedingungen
venture	Unternehmung
work permit *You need a work permit before you can start.*	Arbeitserlaubnis *Sie benötigen eine Arbeitserlaubnis, bevor Sie beginnen können.*

14.5 Übungen – Exercise time

Task 1: What is a/the ...?

1. sending the employee abroad a p________________ or a s________________
2. money the employee receives every month s________________
3. brief listing of the information is an o________________
4. complaints about something at work are g________________
5. young employee with great promise is a h________________ p________________

Task 2: Which word is missing?

1. That's one ________________ why we've come together. (*Grund*).
2. That's right. ________________ we have to do something. (*trotzdem*).
3. ________________ are much higher in Germany than in the UK. (*Abzüge*)
4. There's something I need to get off ________________. (*loswerden*)
5. You'll need to think about ________________ home again in Chicago. (*aufbauen*)

Task 3: Complete the phrases with the missing prepositions.

1. way ________________ life
2. ________________ this reason
3. ________________ my hands
4. ________________ the other hand
5. welcome ________________ headquarters

ARBEITSMITTEL AUF MYBOOK.HAUFE.DE

Auf der Internetseite zum Buch finden Sie weitere Übungen zu diesem Modul. Darüber hinaus bieten wir Ihnen eine Vorlage zur schriftlichen Bestätigung einer Mitarbeiterentsendung an. Nachdem Sie die Namen des Mitarbeiters und die des Unternehmens eingesetzt haben, können Sie diese Bestätigung ausdrucken und verschicken.

15 Präsentationen und Berichte

Zusammenfassung

Mit diesem Modul bieten wir Ihnen zahlreiche nützliche Formulierungen auf Englisch und Hilfen für die Erstellung von Berichten und Präsentationen.

- Mit der *Checkliste* geben Sie Ihrer Präsentation die richtige Struktur und die *Beispielformulierungen* helfen Ihnen, die Inhalte hervorzuheben.
- Tipps und *Musterformulierungen* unterstützen Sie bei der Erstellung eines Kurzberichts und *Textbausteine* beim Verfassen eines aussagekräftigen Personalberichts.

Am Ende des Moduls finden Sie themenspezifische *Vokabeln* und *Übungen*, mit denen Sie Ihr Wissen überprüfen können.

Die Meilensteine: Aussagekräftige Präsentationen halten

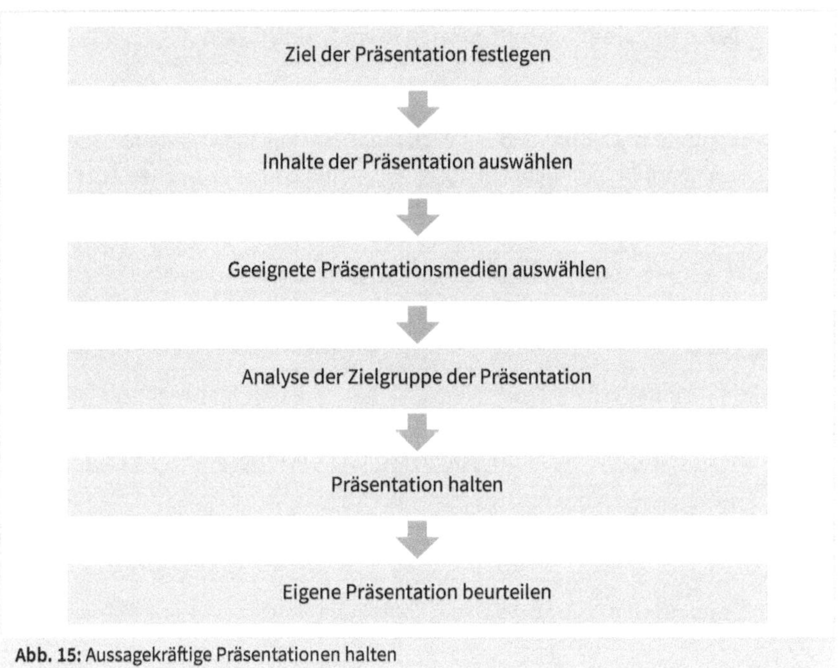

Abb. 15: Aussagekräftige Präsentationen halten

15.1 Präsentationen gekonnt halten

Making a presentation can be a nerve-wracking business for some speakers. The way to manage all this is to be well-organised, know your material very thoroughly and practise it in advance.

Eine Präsentation zu halten, kann für manche Vortragende eine nervenaufreibende Sache sein. Um erfolgreich eine Präsentation zu bewältigen, muss man gut organisiert sein, das vorzutragende Thema durch und durch kennen und vorher üben.

You should prepare different wording for the various parts of a presentation. Below are examples of useful phrases.

Bereiten Sie sich vor, indem Sie die Formulierungen für Ihre Präsentation auswählen. Nachfolgend finden Sie viele solcher Formulierungen.

15.1.1 Textbausteine: Präsentationen klar strukturieren

In diesem Leitfaden geben wir Ihnen griffige Formulierungen an die Hand, mit denen Sie die Präsentation gliedern und die einzelnen Teile einleiten können: Wie Sie sich vorstellen, gekonnt in das Thema einführen, das Publikum ansprechen oder wie Sie die Aufmerksamkeit des Publikums auf sich ziehen.

Introducing yourself – stress things you have in common	**Sich vorstellen, Gemeinsamkeiten betonen**
I'm Markus Maier.	Ich bin Markus Maier.
I spent three years in Britain which is, by the way, where I had this little idea of …	Ich habe drei Jahre in England verbracht, dort ist mir übrigens die Idee gekommen, zu …
I have been a consultant at … for three years.	Ich bin seit drei Jahren als Beraterin bei … tätig.
I know the problems you have to deal with because I too have been a sales manager like most of you.	Ich kenne Ihre Probleme, da ich wie die meisten von Ihnen Verkaufsleiter gewesen bin.
I'm no more expert in this field than you are.	Ich bin auf diesem Gebiet auch nicht mehr Fachmann als Sie.
Thanking your audience	**Bedanken bei den Zuhörern**
I'm glad to have the chance to speak to you about …	Ich freue mich, dass ich die Gelegenheit habe, zu Ihnen über … zu sprechen.
I'm grateful for having been offered the opportunity to present our product range.	Ich bin dankbar, dass man mir Gelegenheit gibt, Ihnen unsere Produktpalette vorzustellen.
I appreciate that you're sacrificing one hour of your valuable time to listen to me.	Ich bin Ihnen dankbar, dass Sie eine Stunde Ihrer wertvollen Zeit opfern, um mir zuzuhören.

Ladies and Gentlemen, I'd like to thank you for your time.	Meine Damen und Herren, ich möchte mich bedanken, dass Sie sich die Zeit genommen haben.
Don't hesitate to interrupt me if you have a question.	Zögern Sie nicht, mich zu unterbrechen, wenn Sie eine Frage haben.
Announcing the topic	**In das Thema einführen**
You've met here today to hear something about our ideas on …	Sie sind heute hier zusammengekommen, um etwas über unsere Vorstellungen von … zu hören.
The topic of this presentation is …	Gegenstand dieser Präsentation ist, …
The subject of today's presentation is …	Das Thema der heutigen Präsentation ist, …
In my presentation I'll be explaining our new project.	In meiner Präsentation werde ich unser neues Projekt erläutern.
What I'm going to explain this afternoon are the problems involved in …	Heute Nachmittag will ich Ihnen die Probleme im Zusammenhang mit … erklären.
I'll be presenting my latest invention.	Ich werde meine neueste Erfindung vorstellen.
This morning I'd like to look back at the past fiscal year.	Heute Vormittag möchte ich einen Rückblick auf das letzte Geschäftsjahr geben.
I'm trying to give you an overview of the development of our company's activities in Asia.	Ich versuche, Ihnen einen Überblick zu geben, wie sich die Aktivitäten unserer Firma in Asien entwickelt haben.
Drawing attention to the handouts	**Aufmerksamkeit auf die Tischvorlagen lenken**
In the handout which I'll be passing round, you'll find …	In den Tischvorlagen, die ich herumgehen lasse, finden Sie …
In the folder which I'll be distributing, you'll find the handouts of my presentation.	In dem Ordner, den ich austeile, finden Sie die Handouts zu meiner Präsentation.
I've prepared a detailed cost breakdown, which I'll be handing round now.	Ich habe eine detaillierte Kostenaufschlüsselung vorbereitet, die ich nun austeilen werde.
As you all know from your pre-conference handouts …	Wie Sie alle Ihren Unterlagen entnommen haben, die Ihnen vor der Besprechung zugegangen sind, …
May I draw your attention to the handout in front of you?	Darf ich Ihre Aufmerksamkeit auf die Tischvorlage vor Ihnen lenken?
Giving an overview of the presentation	**Überblick geben**
My presentation will be in three main parts.	Meine Präsentation gliedert sich in drei Teile.
I'll be developing the three major stages of our project.	Ich werde Ihnen die drei wichtigsten Stufen unseres Projekts schildern
In the first part of my presentation I'll …	Im ersten Teil meiner Präsentation werde ich …

Then, in the second part of the meeting, there will be a period for free discussion and question time.	Im zweiten Teil unserer Besprechung folgt eine Runde für Diskussionen und Fragen.
Firstly, I'd like to … Secondly, we could … And I'll finish with …	Zuerst würde ich gerne … Zweitens könnten wir … Und ich werde schließen mit …
First, I'll give you an overview of … Then I'll … Lastly …	Zuerst gebe ich Ihnen einen Überblick über … Dann werde ich … Schließlich …
Giving some guidance, pointing out steps	**Zusammenfassung ankündigen**
I'll begin by summarising the problems we're having with …	Ich beginne mit einer Zusammenfassung der Probleme, die wir mit … haben.
To begin with, let me say …	Lassen Sie mich einleitend sagen …
Let's start with …	Fangen wir mit … an.
I'd like to begin my presentation by giving you an overview of the project.	Ich würde meiner Präsentation gerne einen Überblick über das Projekt voranstellen. Wenn wir uns nun … zuwenden können.
If I could now turn to … / Let me now move on to …	Lassen Sie mich nun zu … kommen.
So that's the present situation. Now let's look at …	Das ist also die gegenwärtige Situation. Betrachten wir nun …
Let me now come to my other functions.	Lassen Sie mich nun auf die anderen Funktionen zu sprechen kommen.
That completes my overview of …	Damit ist mein Überblick über … abgeschlossen.
Preparing the listeners for visual aids	**Vorgehen klären, Umstände ansprechen**
Well, before I start can everybody see the screen?	Bevor ich beginne, können alle die Leinwand sehen?
If you can move your chairs in a little.	Wenn Sie vielleicht Ihre Stühle ein bisschen zusammenrücken würden.
I'll use the flipchart here for any spontaneous points which might arise.	Ich werde alle Punkte, die spontan auftauchen könnten, auf diesem Flipchart festhalten.
Structuring comments	**Strukturierende Bemerkungen**
First, I'd like to show you several slides to explain the technical details.	Zuerst würde ich Ihnen gerne einige Folien zeigen, um die technischen Details zu klären.
The next slide shows …	Die nächste Folie zeigt …
Now I'll show you …	Nun zeige ich Ihnen …
Let's now look at the figures for …	Betrachten wir nun die Zahlen für …
Getting the audience's attention	**Aufmerksamkeit des Publikums erlangen**
As you can see from this chart …	Wie Sie dem Diagramm entnehmen können, …
Observe the difference between …	Beobachten Sie den Unterschied zwischen …

Notice the movement of the robot.	Beachten Sie die Bewegungen des Roboters.
What is interesting to notice is …	Es ist interessant zu sehen, dass …
It is interesting to notice that …	Es ist interessant zu sehen, dass …
If we look more closely at X …	Wenn wir X genauer betrachten …
Please look at the upper right-hand corner of the diagram.	Betrachten Sie nun die obere rechte Ecke der Grafik.
Inviting questions	**Zu Fragen auffordern**
I'd be happy to answer any questions.	Ich beantworte gerne alle Fragen.
If you've any questions, I'd be pleased to answer them now.	Wenn Sie irgendwelche Fragen haben, würde ich sie jetzt gerne beantworten.
I'd welcome any comments.	Ich würde mich über jeden Ihrer Kommentare freuen.
So, what you're asking is why we …	Sie fragen also, warum wir …
If I understand your question correctly, you would like to know …	Wenn ich Ihre Frage richtig verstehe, wollen Sie wissen …
So, do we plan to increase salaries?	Also, planen wir die Gehälter zu erhöhen?
When you say … do you mean …?	Wenn Sie sagen …, meinen Sie damit …?
I'm sorry, could you please repeat your question?	Entschuldigen Sie, könnten Sie Ihre Frage bitte wiederholen?
Being careful when dealing with criticism	**Mit Kritik vorsichtig umgehen**
Yes, I quite see your point. However, you have to admit that …	Ja, ich verstehe schon, was Sie meinen. Jedoch werden Sie zugeben, dass …
Yes, it's something we've thought about a lot, too, but …	Ja, das ist etwas, worüber wir auch viel nachgedacht haben, aber …
That's an accurate observation. On the other hand, if we consider that …	Das ist eine genaue Beobachtung. Andererseits, wenn wir berücksichtigen, dass …
I know it's difficult to accept this decision, but the evidence is there.	Ich weiß, es ist schwer, diese Entscheidung zu akzeptieren, aber die Beweise liegen auf der Hand.
Avoiding criticism and questions	**Kritik und Fragen vermeiden**
I'm afraid I'm not the right person to answer that.	Es tut mir leid, aber ich bin nicht der richtige Ansprechpartner für Ihre Fragen.
Dr XY is a much better person to answer that.	Dr. XY kann diese Frage viel besser beantworten.
I'm not sure this is the right place to discuss this particular question.	Ich bin mir nicht sicher, ob dies der richtige Ort ist, um diese spezifische Frage zu erörtern.
Could we leave that for later?	Könnten wir dies auf einen späteren Zeitpunkt verschieben?

How to say goodbye	Gekonnt verabschieden
If you should have any further questions feel free to phone me.	Sollten Sie noch irgendwelche Fragen haben, können Sie mich ruhig anrufen.
Thank you for your attention, Ladies and Gentlemen.	Vielen Dank für Ihre Aufmerksamkeit, meine Damen und Herren.
So that was it for today. Thank you for your attention.	So, das war's für heute. Vielen Dank für Ihre Aufmerksamkeit.
Thank you for coming. I hope you enjoyed it.	Danke, dass Sie gekommen sind. Ich hoffe, es hat Ihnen gefallen.

15.1.2 Textbausteine: Inhalte der Präsentation hervorheben

Beispielhafte Formulierungen, um die Aufmerksamkeit auf bestimmte Inhalte der Präsentation zu lenken, finden Sie in dieser Tabelle.

Referring to visuals	Auf visuelle Hilfsmittel hinweisen
This shows/illustrates/demonstrates/refers to …	Dieses zeigt/illustriert/bezieht sich auf …
This is a graph/an organigram which shows …	Dies ist eine Grafik/ein Organigramm welches … zeigt.
As you can see, this is …	Wie Sie sehen können, ist dies …
Here we can see …	Hier können wir … sehen.
Focusing attention	**Aufmerksamkeit auf etwas lenken**
I'd like to draw your attention to …	Ich würde gerne Ihre Aufmerksamkeit auf … lenken.
One of the most important aspects of this is …	Dies ist einer der wichtigsten Aspekte von …
At first glance it seems … but …	Oberflächlich betrachtet scheint es … zu sein, aber …
Talking about charts and figures	**Über Grafiken und Zahlen sprechen**
This curve represents …	Diese Kurve zeigt/stellt … dar.
Let's now look at the figures for …	Werfen wir nun einen Blick auf Zahlen für …

If we compare these two curves then …	Wenn wir diese beiden Kurven vergleichen, dann …
To reach the peak	Den Höchststand erreichen
To hit rock bottom/reach the lowest point	Die Talsohle erreichen
Notice the difference between …	Beobachten Sie den Unterschied zwischen …
It is interesting to notice that …	Es ist interessant zu sehen, dass …
If we look more closely at …	Wenn wir … genauer betrachten
Please look at the upper right-hand corner of the graphic.	Schauen Sie nun auf die obere rechte Ecke der Grafik.
This chart compares turnover in the two countries.	Dieses Diagramm vergleicht die Umsätze in den beiden Ländern.
So you will, no doubt, see instantly that …	Sie werden sicherlich sofort erkennen, dass …
Here you can see the development over the past two years.	Sie können hier die Entwicklung der letzten beiden Jahre verfolgen.

15.1.3 Checkliste: Präsentationen einfach gestalten

Ein typischer Fehler bei Präsentationen ist, dass Präsentationsfolien mit Informationen überladen sind. Der Trick ist es, alles so einfach wie möglich zu halten und die Folie so übersichtlich zu gestalten, dass das Publikum sie rasch nachvollziehen kann.

Keep the following in mind:	**Beachten Sie folgende Punkte:**
Keep the slides simple.	Gestalten Sie die Folien einfach.
All parts of the slide must be easily readable.	Alle Teile der Folie müssen einfach zu lesen sein.
Use bullet points for each item.	Benutzen Sie Aufzählungszeichen für jeden einzelnen Punkt.
Don't use more than five bullet points.	Benutzen Sie nicht mehr als fünf Aufzählungszeichen.
List the information in logical order.	Führen Sie die Informationen in einer logischen Reihenfolge auf.

15.2 Die Gliederung des Berichts übersichtlich gestalten

Short report-writing is often needed when a business situation arises which is out of the ordinary.

Die Fähigkeit, kurze Berichte zu schreiben, ist von Nutzen, wenn eine geschäftliche Situation außergewöhnliche Formen annimmt.

For example, if negotiations or the quality of a new product is unacceptable to the boss, he or she can delegate the task of researching the situation and compiling a report about it with recommendations for action.

Wenn zum Beispiel Verhandlungen oder die Qualität eines neuen Produkts für die Führungskraftnicht akzeptabel ist, kann er oder sie die Aufgabe an einen:e Mitarbeiter:in delegieren und darum bitten, die Situation zu untersuchen und einen Bericht darüber zu verfassen, in dem auch Empfehlungen für das weitere Vorgehen angeführt werden.

15.2.1 Übersicht: Struktur eines Berichts

There are various ways to set out a report. One simple layout useful in many cases is the following:	Es gibt verschiedene Möglichkeiten, einen solchen Bericht zu verfassen. Einen einfachen und für viele Situationen anwendbaren Aufbau stellen wir Ihnen im Folgenden vor.

Terms of reference	**Aufgabenstellung**
Specify subject of report, who has commissioned the report and date of report.	Nennen Sie den Gegenstand des Berichts, die Persion, die den Bericht in Auftrag gebeben hat sowie das Erstellungsdatum.
Proceedings	**Vorgehensweisen**
Describe what actions were taken to gather the facts.	Beschreiben Sie, was unternommen wurde, um die Fakten zu ermitteln.
Results	**Ergebnisse**
Specify what findings were made.	Geben Sie an, welche Ergebnisse erzielt wurden.
Conclusions	**Schlussfolgerungen**
Specify the conclusions drawn from the results.	Listen Sie auf, welche Schlussfolgerungen Sie aus den Ergebnissen gezogen haben.
Recommendations	**Empfehlungen**
Make practical suggestions as to what should be done.	Geben Sie praktische Empfehlungen, was getan werden kann.

15.2.2 Beispielbericht: Report on the late arrival of some office staff

Background: The office manager has noticed that staff are arriving increasingly late during the last three months. He needs an employee to investigate and possibly recommend action.

Report on the late arrival of some office staff over the last three months
Terms of reference

Report on the increase of late arrival of some of the office workers over the last three months. Commissioned by Head of Purchasing Department.

Proceedings

All members of the purchasing department were questioned individually on the sudden increase in the lack of punctuality.

Results

The latecomers are those who travel to work by private transport and public road transport. Those who travel by train arrive on time.

The two main roads leading into town have had major road works done on them for the last twelve weeks; this causes unavoidable traffic jams and long delays.

About three months ago our company opened its new mail order section. About forty new employees were recruited. As this new section starts work forty-five minutes before the employees in the purchasing department, the car park gets filled with vehicles belonging to workers in that section. The company has not enlarged the car park to accommodate the additional vehicles. Consequently, many office staff cannot find parking spaces and have to drive around surrounding roads looking for a space to park.

The above information was gathered from my individual interviews with staff.

Conclusions

Clearly, nothing can be done about the road works. It is understood that it will be finished in six weeks' time.

As people will insist on travelling to work in their own cars, the best solution for the company would be to either enlarge the existing car park or to acquire another one near the office.

Recommendation

As the road works will be finished in six weeks, staff should be asked to leave home for work a little earlier than usual for this period to enable them to arrive on time. It is advisable that the company enlarges the existing car park as it is surrounded by company land.

[Signature] [Date]

HR Helpdesk Team Manager

15.3 Personalberichte verfassen

It is advisable to use spreadsheet applications for reports on facts and figures on employees in divisions, departments or the company as a whole.

Der Einsatz von Tabellenkalkulationsprogrammen empfiehlt sich, um Berichte mit Fakten und Zahlen zur Personalsituation in Abteilungen, Niederlassungen oder dem gesamten Unternehmen zu verfassen.

15.3.1 Gliederung eines Personalberichts

The topics of the individual spreadsheets could be as follows:

Sheet 1	Personal details of employees
Sheet 2	Salary structure
Sheet 3	New recruits
Sheet 4	Resigned employees
Sheet 5	Other reports

Die Gliederung für die einzelnen Arbeitsblätter kann folgendermaßen aussehen:

Blatt 1:	Persönliche Mitarbeiterdaten
Blatt 2:	Gehaltsstruktur
Blatt 3:	Neueinstellungen
Blatt 4:	Ausgeschiedene Mitarbeiter
Blatt 5:	Andere Berichte

15.3.2 Textbausteine: Aussagekräftige Personalberichte verfassen

Nachfolgend stellen wir Ihnen Musterformulierungen für die Erstellung eines Personalberichts vor.

Mentioning HR events	HR Ereignisse erwähnen
Awards	**Auszeichnungen**
HR Director received the XY Award.	Der Personalleiter erhielt die XY Auszeichnung.
HR Director was awarded the XY Competence Award.	Die Personalleiterin wurde mit dem XY Preis ausgezeichnet.
Sponsoring	**Förderung**
EURXX to sponsor XY.	XX EUR für das Sponsoring von XY.
An investment of EUR XX to support …	Eine Investition von XX EUR zur Unterstützung von …
XY will receive/get a sponsorship of EUR XX.	XY erhält eine Förderung von XX EUR.
Further education	**Fortbildung**
Employees were given the opportunity to study English.	Den Mitarbeitenden wurde die Gelegenheit gegeben, Englisch zu lernen.
Employees get the chance to learn English.	Mitarbeiternde erhalten die Möglichkeit, Englisch zu lernen.
Chance of learning English for employees.	Möglichkeit für Angestellte, Englisch zu lernen.
Donations	**Spenden**
XY donated EUR XX to XYZ.	XY spendete XX EUR für XYZ.
EURXX was given to XY as aid.	XX EUR wurde an XY als Unterstützung gezahlt.
A donation of EURXX was made to XY.	Eine Spende in Höhe von XX EUR wurde an XY gezahlt.
Employees	**Mitarbeiter**
XY constantly works to improve and adapt the recruitment process.	XY arbeitet permanent an der Verbesserung und Anpassung des Einstellungsprozesses.
XY works to enhance and adapt the recruitment process.	XY arbeitet ständig an der Verbesserung und Anpassung des Einstellungsprozesses.
The recruitment process is being constantly/continuously improved and adapted by …	Der Einstellungsprozess ist konstant durch … verbessert und angepasst worden.

Mentioning HR events	HR Ereignisse erwähnen
Workforce	**Belegschaft**
At present XY employs a workforce of XX.	Derzeit beschäftigt XY XX Mitarbeitende.
The number of employees has risen from XX to XX.	Die Mitarbeiterzahl ist von XX auf XX gestiegen.
In the period under review, the company employed xx new recruits.	In Berichtszeitraum hat das Unternehmen XX neue Mitarbeiter eingestellt.
Redundancies	**Entlassungen**
XX employees were made redundant.	XX Mitarbeiter sind entlassen worden.
XX employees will lose their jobs.	XX Mitarbeiter werden ihre Arbeitsstelle verlieren.
XX jobs axed at XY.	XX Arbeitsplätze wurden in XY abgebaut.
Due to the current economic downturn we had to let XX employees go.	Aufgrund des gegenwärtigen wirtschaftlichen Abschwungs mussten wir XX Mitarbeiter entlassen.
Skills	**Kompetenzen/Fähigkeiten**
It is vital for all of us to acquire a broader selection of skills.	Es ist für uns alle unerlässlich, breiter gefächerte Fähigkeiten zu erwerben.
We must all see skills in a wider sense.	Wir müssen Fähigkeiten in einem weiteren Sinn sehen.
It is absolutely necessary for everybody to acquire a larger skills spectrum.	Es ist für jeden unumgänglich, eine größere Bandbreite von Fähigkeiten zu erlangen.
Communication	**Kommunikation**
Communication between people, companies and society is vital for life and business to run smoothly.	Die Kommunikation zwischen Menschen, Firmen und innerhalb der Gesellschaft ist für ein reibungslos verlaufendes Arbeits- und Privatleben unerlässlich.
People, companies and society must communicate with each other if life and business are to run smoothly.	Wenn Arbeits- und Privatleben reibungslos laufen sollen, müssen Menschen, Firmen und Gesellschaft miteinander kommunizieren.
If we want life and business to run smoothly, then people, companies and society have to communicate with one another.	Wenn wir wollen, dass das Arbeits- und Privatleben. reibungslos laufen, müssen Menschen, Firmen und Gesellschaft miteinander kommunizieren.

Mentioning HR events	HR Ereignisse erwähnen
Working environment	**Arbeitsumfeld**
At XY, we see organisational efficiency, good leadership and a favourable psycho-social environment as central factors.	Bei XY sehen wir organisatorische Effizienz, gute Mitarbeiterführung und ein erfolgreiches psycho-soziales Umfeld als zentrale Faktoren an.
At XY, we consider the efficiency of the organisation, good leadership and a positive mental environment as key.	Bei XY halten wir die organisatorische Effizienz, gute Mitarbeiterführung und ein positives mentales Umfeld als für entscheidend.
Organisational efficiency, good leadership and a psycho-social environment are seen/considered as key.	Organisatorische Effizienz, gute Mitarbeiterführung und ein psycho-soziales Umfeld werden als entscheidend angesehen.
Safety	**Sicherheit**
The top priority areas for skills development in the last 12 months were working environment, risk management and safety.	Die vordringlichsten Bereiche der Potenzialentwicklung im letzten Jahr waren Arbeitsumfeld, Risikomanagement und Sicherheit.
Skills that had to be improved over the last year were working environment, risk management and safety.	Fähigkeiten, die im letzten Jahr verbessert werden mussten, waren das Arbeitsumfeld, Risikomanagement und Sicherheit.
Top of the list for skills development over the last year were the areas of working environment, risk management and safety.	Ganz oben auf der Liste für Potenzialentwicklung standen im letzten Jahr die Bereiche Arbeitsumfeld, Risikomanagement und Sicherheit.

15.3.3 Beispielbericht: Example of a HR report

HR events in [Year]
A master of skills development

HR Director [Name] received the Competence Award for Human Resources Directors at the 2009 Competence Gala for his structured efforts in uniting personnel and business development.

[Company] makes [Number] employees redundant

The news came at the end of March that [Number] employees within the Group's chemicals division would be made redundant.

EUR [Amount] donated to [Name of social institution]

[Company name] donated EUR [Amount] to [Name of institution]. This sponsorship is in accordance with the Group's Code of Conduct.

Study English through interactive training

In September, [Company name]'s employees were given the opportunity to study English in an interactive internet-based training scheme. Several employees joined the scheme, which is being implemented in [Year].

Key figures 2009
Average number of employees

	2009		2008	
Country	**Total number of employees**	**of whom men**	**Total number of employees**	**of whom men**
Belgium	207	185	212	189
France	7	5	11	6
Germany	138	123	145	128
India	232	225	241	233
Italy	92	80	99	87
Japan	2	1	2	1
Netherlands	1	1	2	2
Singapore	4	2	4	2
South Korea	28	25	15	14
UK	5	3	8	6
USA	310	271	312	275
Total	**1,026**	**921**	**1,051**	**943**

Employees
Improved recruitment

[Company] constantly works to improve and adapt the recruitment process. This work is based on constant evaluation of our recruiting procedures. Our company policy states that all appointments should be advertised internally. In 2009 a change occurred in the digital advertising procedures with the introduction of a new tool for personnel administration. The Group's website, and other external search methods have been used for external recruitment.

Skills
Constant development

It is vital for all of us to adopt a wider view of skills development that includes more than just traditional training methods. Most opportunities to learn from each other occur in everyday life, in situations such as work rotation, trainee work, project participation or study visits. Working on development plans in a structured way allows us to make the most of these opportunities.

Communication
Communication for achieving goals

Communication is a strategically important management tool. Communication is integrated into the Group's business, so it is not regarded as an isolated activity.

Working environment/health
Health means both physical and mental well-being

The World Health Organization's definition of health is: »The greatest possible well-being of the individual, physical, spiritual, intellectual and social, and not just in the absence of illness.«

We see organisational efficiency, good leadership and a favourable social environment as central factors for employee health and well-being.

Staff satisfaction is an indicator of how our employees feel.

Safety

Safety at [Company] is a priority and not only in manufacturing where risk factors are highest. Skills development in the year under review was focused on working environment, risk management and safety. Accident prevention measures are implemented at all our sites around the world.

15.4 Vokabeln – Vocabulary input

Nachfolgend finden Sie eine Auflistung wichtiger Fachbegriffe aus diesem Modul.

English	German
accident prevention	Unfallverhütung
accommodate	unterbringen
accurate *The chairperson requires accurate figures.*	genau *Der:Die Vorstandsvorsitzender:e benötigt genaue Zahlen.*
appreciate	dankbar sein
audience	Publikum
award	auszeichnen
break-down	Aufschlüsselung
bullet point	Aufzählungspunkt
code of conduct	Verhaltenskodex
common; to have something in ~	etwas gemeinsam haben
compile *Please compile all data into a report.*	zusammenstellen *Bitte stellen Sie alle Daten zu einem Bericht zusammen.*
conclusion	Schlussfolgerung
curve *This curve shows a clear upward trend.*	Kurve *Diese Kurve zeigt einen eindeutigen Trend nach oben.*
decision	Entscheidung
distribute *Please distribute the new report to all employees.*	verteilen *Bitte verteilen Sie den neuen Bericht an die gesamte Belegschaft.*
donate	spenden
draw attention to something	Aufmerksamkeit auf etwas lenken
economic downturn	Konjunkturabschwung, wirtschaftlicher Rückgang
enhance	verbessern
enlarge	vergrößern
evidence	Beweis
findings	Erkenntnisse

English	German
folder	Ordner; Mappe
gather	sammeln
grasp something *She explained the concept in an easy-to-grasp way.*	etwas verstehen, begreifen *Sie erläuterte das Konzept auf eine einfach zu verstehende Weise.*
guidance *Without guidance from the management we're finished.*	Führung; Anleitung *Ohne Führung durch das Management sind wir erledigt.*
handout	Handout, Tischvorlage, Informationsblatt
hesitate *Don't hesitate to ask.*	zögern *Zögern Sie nicht zu fragen.*
invention	Erfindung
leadership	Mitarbeiterführung
nerve-wracking	nervenaufreibend
observation	Beobachtung
overload	überladen
peak; reach a ~	einen Höchststand erreichen
proceedings	Vorgehensweisen
projected image *The projected images show you our profit for the year.*	dargestellte o. gezeigte Bilder *Die dargestellten Bilder zeigen Ihnen unseren Gewinn für das Jahr.*
punctuality	Pünktlichkeit
recommendation	Empfehlung
redundant; make someone ~ *30 employees were made redundant due to lack of orders.*	aus betriebsbedingten Gründen entlassen *30 Arbeitskräfte wurden aufgrund fehlender Aufträge entlassen.*
rock bottom; hit ~	Talsohle erreichen
sacrifice	opfern
satisfaction	Zufriedenheit
screen	Bildschirm; Leinwand
self-explanatory	selbsterklärend
slide	Folie
sponsorship	Förderung

English	German
spreadsheet	Tabellenkalkulation
summarise	zusammenfassen
surround *We are surrounded by positive thinking.*	umgeben *Wir sind von positiven Gedanken umgeben.*
terms of reference	Aufgabenstellung
thorough	gründlich
timing	zeitliche Planung
transparency	Folie
well-being	Wohlbefinden
working environment *We enjoy a pleasant working environment.*	Arbeitsumfeld *Wir genießen ein angenehmes Arbeitsumfeld.*

15.5 Übungen – Exercise time

Task 1: Find a word in the texts which has the same meaning as the following (underlined).

1. He should <u>investigate</u> whether … ________________
2. I'm no <u>specialist</u>. ________________
3. To offer the <u>chance</u> ________________
4. That <u>finishes</u> my overview. ________________
5. The upper right-hand corner of the <u>chart</u> ________________
6. help or <u>support</u> ________________

Task 2: Vocabulary expansion: Fill in the missing words.

verb	noun	adjective	German
guide			
visualise			
	comparison		
introduce			
find			
		punctual	
			spenden
		continuous	

Task 3: What is the English for ...

1. Ich bin seit drei Jahren als Beraterin bei ICS tätig.
2. Heute will ich einen Rückblick auf das letzte Geschäftsjahr geben.
3. Lassen Sie mich einleitend sagen, dass ...
4. Ich beantworte gerne alle Fragen.
5. Darf ich Ihre Aufmerksamkeit auf die Tischvorlage vor Ihnen lenken?

ARBEITSMITTEL AUF MYBOOK.HAUFE.DE

Auf der Internetseite zum Buch finden Sie weitere Übungen zu diesem Modul sowie zahlreiche Textbausteine für Ihre Präsentation.

16 Reden zu Firmenanlässen

Zusammenfassung

Mit den richtigen Worten kann man vieles bewegen. Wir haben in diesem Modul für Sie praktische *Arbeitsmittel* und zahlreiche *Musterformulierungen* zusammengestellt, um Sie bei der Vorbereitung zu unterstützen und Sie für Ihre Rede zu inspirieren. Wir gehen auf verschiedene Anlässe ein und bieten Ihnen sowohl allgemeinere *Beispielformulierungen*, aber auch *Textbausteine* für eine Motivationsrede und auch für den traurigen Fall, dass Sie Worte des Beileids zu übermitteln haben.

Am Ende des Moduls haben wir die wichtigen *Vokabeln* für Sie alphabetisch aufgelistet und drei *Aufgaben* notiert, mit denen Sie Ihr Wissen überprüfen können.

Die Meilensteine: Eine wirkungsvolle Rede halten

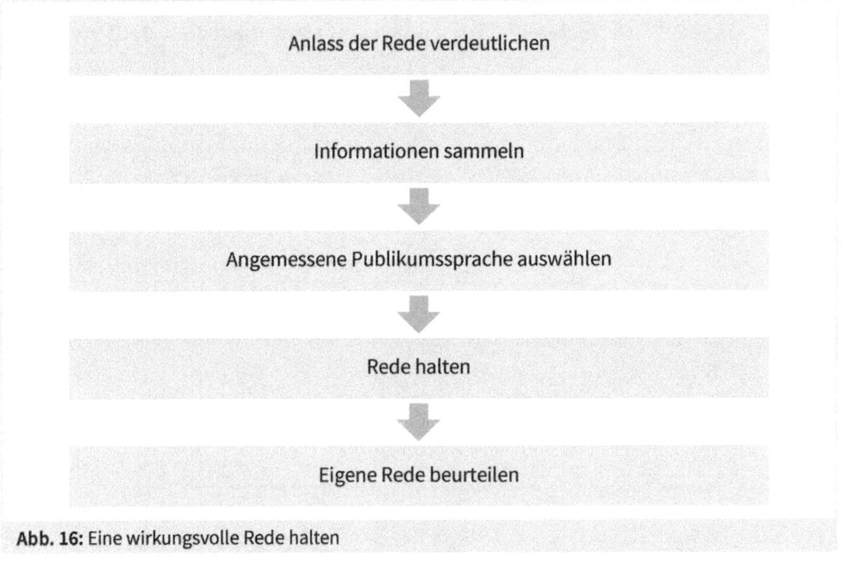

Abb. 16: Eine wirkungsvolle Rede halten

16.1 Eine Rede im Unternehmen halten

There are many occasions for saying a few words or even making longer speeches in the company.
Such speakers are often department leaders or executives. However members of HR are also frequently charged with these tasks.

Occasions for speeches, proposing a toast or simply saying a few words are:

- reviews looking back on how the past year was,
- a few words toasting the prosperity of the company in general,
- praising the commitment of staff,
- replying to the toast of a business partner,
- celebrating long years of service of employees,
- and words of condolence on the death of an employee or a person that was close to the company.

So that you are not short of the right words you find many phrases below.

Es gibt viele Gelegenheiten, in der Firma ein paar Worte zu sagen oder sogar eine Rede zu halten.
Häufig ist es die Aufgabe der Abteilungsleiter:innen oder der Führungskräfte, Reden zu halten. Oft werden aber auch Mitarbeiter:innen aus der Personalabteilung damit betraut.

Anlässe, eine Rede zu halten, einen Toast auszubringen oder schlicht ein paar Worte zu sagen, sind:

- der Rückblick auf das vergangene Jahr,
- Worte auf das Wohl der Firma im Allgemeinen,
- das Engagement der Mitarbeitenden zu loben,
- den Toast eines Geschäftspartners zu erwidern,
- langjährige Betriebszugehörigkeiten von Mitarbeitenden zu feiern,
- Worte der Anteilnahme zum Gedenken an verstorbene Mitarbeitende oder dem dem Unternehmen nahestehende Personen.

Damit Ihnen nicht die passenden Worte fehlen, bieten wir Ihnen im Folgenden zahlreiche Formulierungen an.

16.1.1 Textbausteine: Langjährige Firmenzugehörigkeit würdigen

Die richtige Rede bei einem Mitarbeiterjubiläum zu halten, ist gar nicht so einfach. Nachfolgend finden Sie ein paar hilfreiche Musterformulierungen für diesen Anlass.

Introducing the event, the person	Anlass und Thema beschreiben
In our company it's a custom to say a few words on a pleasant occasion.	In unserer Firma ist es ein Brauch, bei einem erfreulichen Anlass ein paar Worte zu sagen.
It's an opportunity to say what we've all been thinking.	Es ist die Gelegenheit, zu sagen, woran wir alle gedacht haben.

I'm very happy to have the chance to say a few words about …	Ich bin sehr glücklich, die Gelegenheit zu haben, einige Worte über … zu sagen.
Now's the chance to say a big »thank you« for …	Nun ist die Gelegenheit, ein großes Dankeschön für … zu sagen.
Remembering the past	**Vergangenes erinnern**
I remember one occasion in particular when …	Ich kann mich an eine einzelne Begebenheit erinnern, als …
Thinking back there was the situation when …	Zurückblickend gab es die Situation, als …
I don't know if you can still remember it but …	Ich weiß nicht, ob Sie sich daran erinnern können, aber …
Well, I'm sure we all still have in mind …	Nun, ich bin sicher, dass wir uns alle immer noch an … erinnern.
Going into detail	**Details benennen**
If you'll allow me, I'd like to mention …	Mit Ihrer Erlaubnis würde ich gerne … erwähnen.
I'd like to pick out one detail. On one occasion you …	Ich würde gerne ein Detail herausgreifen. Einmal haben Sie …
There was one thing that stood out. It was years ago on …	Da gab es eine herausragende Sache. Vor einigen Jahren am …
Let's just focus on one detail. We …	Konzentrieren wir uns auf eine Einzelheit. Wir …
Talking about character	**Charaktereigenschaften beschreiben**
Compassion is what guided you in your work.	Ihre Arbeit wurde stets von Menschlichkeit geleitet.
I'm sure we all know how important fair play is to you.	Ich bin sicher, dass wir alle wissen, wie wichtig Fair Play für Sie ist.
We all know you were always driven by team spirit.	Wie wir alle wissen, waren Sie immer getrieben vom Gedanken des Teamgeists.
Toasting the employee	**Mitarbeiter:innen hochleben lassen**
So, let me raise my glass to you, [Name].	Nun, lassen Sie mich mein Glas auf Sie, [Name], erheben.
Now, I'd like to toast [Name].	Nun würde ich gerne einen Toast auf [Name] ausbringen.
Please all rise to toast [Name].	Erheben Sie sich bitte alle, um einen Toast für [Name] auszubringen.
Let's all raise our glasses to [Name].	Lassen Sie uns unsere Gläser für [Name] erheben.

16.1.2 Checkliste: Den Redeneinstieg gestalten

Tips on what to consider for the opening of speeches:

Compliment entry	
Greet the audience	Begrüßung des Publikums
Address honorary and important guests by name and title	Honoratioren und wichtige Gäste mit Namen und Titel begrüßen
Create a feeling of belonging	**Gefühl der Zusammengehörigkeit**
Highlight shared values	Verbindende Werte herausstellen
Point out common interests	Gemeinsame Interessen aufzeigen
Introducing the event	**Den Anlass beschreiben**
Explain why you have come together	Erklären, warum man zusammengekommen ist
Introduce the subject	Das Thema umreißen
Short historical summary	**Kurzer Rückblick**
What we achieved together	Was wurde zusammen erreicht
Give an optimistic outlook to the future	Geben Sie einen optimistischen Ausblick auf die Zukunft

16.1.3 Beispielrede: Speech celebrating an employee's years of service

Dear Rachel,

Dear Colleagues,

In our company it is our custom to say a few words on a pleasant occasion like this. Rachel, as your boss I'd like to say something on this occasion. And I want to start by stressing that it's a wonderful chance to say those two words that you have heard so often from many people over the years: »Thank you!« Thank you for being with us for 25 years now.

I remember one occasion in particular. It happened about 23 years ago when we were running out of time in a very important project. You all remember the »Munich Miracle« as it was later recorded in the chronicles of this company. At that time we were developing a new product. And that was necessary! Our competitors were moving in the same direction. Of course, as the market leader we had to be the first. Rachel, you were only a junior team member then but you made sure that we stayed on track and on time. You were tenacious and determined. You pushed for results and at times, when I, as the project leader was feeling desperate, you even pushed me. And I am glad you did.

If you'll permit I'd like to mention one more character trait: compassion. As I said, you stood by your team when times were rough. And you even stood in for colleagues when they had real problems. Each and every one of us here tonight knows that and has had that kind of support from you at some time in the past 25 years.

Compassion is also what guided you in your main hobby. Once a month you volunteer for work in a soup kitchen for homeless people. You spend time with the poorest of the poor and do so as naturally and gracefully as you work with all of us. Sometimes you seem to burn the candle at both ends. Yet I know: you have more than enough energy and also have a family that help you recharge your batteries.

So let me raise my glass and toast you, Rachel. As I said: Thank you! Thank you for putting in the extra effort because good has never been good enough for you. Thank you for being a trusted and respected co-worker. Thank you for being with us!

16.1.4 Checkliste: Eine erfolgreiche Rede halten

Was Sie beim Schreiben und Vortragen einer Rede alles beachten sollten, erfahren Sie hier.

Be prepared	
• Do I know my audience? Who are they? What do they expect? • What do I expect them to remember or do? • Have I got a clear outline of my speech? • Is it clearly structured? • Have I cut out all the less important facts? • Have I included enough examples to support my arguments? • Shall I use OHP, flipchart, digital projector, handouts etc? • Do I have a provocative, ironic opening to my speech? • How will I end my speech? • How do I dress for success? Are my clothes adapted to the age and position of my listeners?	

Good intentions	
• I'll try to identify with my audience. • I won't speak too fast. • I won't read my speech. • I'll speak loudly and clearly. • I'll keep my body language under control • I'll remember the pauses where necessary. • I'll keep eye contact with my adudience. • I won't let them provoke me. • I'll answer even the most stupid question politely. • I'll think positive and be optimistic whatever happens.	
Last minute check at the set	
• Do I need a microphone? • Should I use the rostrum or not? • Can everybody see me? • Do the OHP and the digital projector function properly? • Can everybody see the screen?	

16.2 Motivationsreden

Company celebrations represent a great opportunity to thank the staff and thus to motivate them. The speech should include some positives, some successes of the employees.

Generally you begin with thanking the staff, then give some detail about the work itself in a positive way and teamwork involved and at the end make a final toast.

Firmenfeiern bieten eine großartige Gelegenheit, der Belegschaft zu danken und sie dadurch auch zu motivieren. Die Ansprache sollte einiges Positives sowie ein paar Erfolge der Belegschaft beinhalten.

Üblicherweise sollten Sie mit einem Dankeschön an die Belegschaft beginnen und danach mit einigen positiven Einzelheiten über die eigentliche Arbeit und der damit einhergehenden Zusammenarbeit weitermachen, enden sollten Sie mit einem abschließenden Trinkspruch.

16.2.1 Textbausteine: Eine motivierende Rede halten

Saying thank-you	Dank sagen
It is my privilege and pleasure to thank you all for your work …	Es ist ein Privileg und mein Vergnügen, mich bei Ihnen für Ihre Arbeit zu bedanken.
On behalf of the management, I would like to thank you for your commitment.	Im Namen der Geschäftsleitung möchte ich mich bei Ihnen für Ihren Einsatz bedanken.
The reason for our gathering today is to wish all the best to …	Der Grund unseres heutigen Zusammenkommens ist, … alles Gute zu wünschen.
I want to grasp the opportunity to say thank you to you all.	Ich möchte die Gelegenheit ergreifen, mich bei Ihnen allen zu bedanken.
Some detail on work itself	**Sinn der Arbeit**
Personally I think work is a mixture of enjoyment and a necessary evil.	Persönlich denke ich, dass Arbeit eine Mischung aus Vergnügen und notwendigem Übel ist.
Well, work is something we all need to do, but …	Nun, Arbeit ist etwas, was wir alle tun müssen, aber …
Work can be done in a spirit of mutual cooperation.	Arbeit kann im Geist der Zusammenarbeit leichter bewältigt werden.
The question is »do we work to live or live to work«?	Die Frage ist, ob wir arbeiten, um zu leben oder Leben, um zu arbeiten?
Talking about teamwork	**Über Teamarbeit reden**
Teamwork means loyalty and I know that no staff could be more loyal.	Zusammenarbeit bedeutet Loyalität und ich weiß, dass keine Belegschaft loyaler sein kann.
Teamwork is the key to our success.	Zusammenarbeit ist der Schlüssel zu unserem Erfolg.
Without teamwork we are nothing.	Ohne Teamarbeit sind wir nichts.
I want you to know your loyalty is appreciated.	Ich möchte, dass Sie wissen, dass Ihre Loyalität geschätzt wird.
Final toast	**Abschließender Gruß**
Ladies and Gentlemen, I've talked enough. I would like to propose a toast to the health of our staff.	Meine Damen und Herren, ich habe genug gesprochen. Lassen Sie mich auf die Gesundheit unserer Belegschaft anstoßen.
Okay, that was enough. Here's to the health of our staff.	Okay, das war genug. Ich trinke auf das Wohl unserer Mitarbeitenden!
Let's raise our glasses to the prosperity of our staff and our company.	Lassen Sie uns das Glas erheben zum Wohl unserer Belegschaft und unserer Firma.
Okay everybody, those were enough words. Now a toast to our staff!	Okay, alle miteinander, das waren genug Worte. Lassen Sie uns auf unsere Belegschaft anstoßen.

16.2.2 Beispielrede: Speech addressing teamwork and loyalty

Ladies and Gentlemen,

It is my privilege and pleasure to propose a toast to the health of the staff of R&D. The reason why it is a pleasure is that I welcome the opportunity to say a few things that I think every day. Whether work is a positive source of enjoyment, or merely a necessary evil, is a matter of opinion – personally I think it is a mixture of both; but there can be no doubt that it is a good deal more pleasant – or, if you like, a good deal less unpleasant – if it is done in a spirit of mutual co-operation. We have that spirit here to a very high degree. I hope we shall always keep it.

Now I am not going to make a high-flying speech about working for the honour and glory of the company. This is a business, and we are in business to make money; and you all work in the company with the same basic objective. But, as you have shown, there is no reason why even hard work should not be made pleasant; and the way to make it pleasant is to work as a team. Teamwork means loyalty, and I know that no staff could be more loyal. I want you to know that your loyalty is appreciated.

I have talked enough. Ladies and Gentlemen, the health of the staff!

16.3 Beileidsworte und Kondolenzschreiben

This is, of course, a very difficult time. The people concerned do not normally wish to hear long drawn-out speeches. It's enough to say the appropriate sentence. Below are some typical examples.

Das ist natürlich eine sehr schwierige Zeit. Die Betroffenen möchten normalerweise keine langatmigen Reden hören. Häufig ist es besser, einige wenige Sätze zu sagen. Nachfolgend ein paar typische Beispiele.

16.3.1 Bespielformulierungen: Beileid in angemessenen Worten ausdrücken

Dieser Leitfaden hilft Ihnen, im Trauerfall die richtigen Worte zu finden.

Expressing condolences	Beileid bekunden
Please accept my deepest sympathy.	Ich möchte Ihnen mein tiefstes Mitgefühl aussprechen.
I don't know what to say.	Ich weiß nicht, was ich sagen soll.
It's a great loss to all of us.	Es ist ein großer Verlust für uns alle.

I'm so sorry to hear about your husband's death.	Es tut mir so leid von dem Tod Ihres Mannes zu erfahren.
I'm sorry to hear that XY has passed away.	Es tut mir leid zu hören, dass XY verstorben ist.
I heard that XY is no longer with us. My deepest sympathy.	Ich hörte, dass XY nicht mehr unter uns ist. Mein tiefstes Mitgefühl.
Her passing away is a great loss to all of us.	Ihr Tod ist ein großer Verlust für uns alle.
I'd like to express my condolences.	Ich möchte Ihnen mein herzliches Beileid aussprechen.
On behalf of the company I'd like to express my condolences.	Im Namen der Firma möchte ich Ihnen mein Beileid aussprechen.
The whole department feels with you on this.	Die ganze Abteilung fühlt mit Ihnen.
We all feel with you on this.	Wir alle fühlen mit Ihnen.
Can I offer you my condolences?	Darf ich Ihnen mein Beileid aussprechen?
Offering support	**Hilfe anbieten**
Let me also offer my support and assistance during this difficult time.	Lassen Sie mich auch meine Hilfe und Unterstützung in dieser schwierigen Zeit anbieten.
If I can do anything to help.	Wenn ich irgendwas tun kann, um zu helfen.
Please let me know, if there's anything I can do.	Lassen Sie es mich wissen, wenn es etwas gibt, was ich tun kann.
Can I be of help in any way?	Kann ich in irgendeiner Weise helfen?
Is there anything I or the company can do for you?	Gibt es irgendwas, was ich oder die Firma für Sie tun kann?
Asking if the person is okay	**Nach dem Befinden fragen**
Are you sure you're okay?	Sind Sie sicher, dass es Ihnen gut geht?
You're sure you're managing alright?	Sind Sie sicher, dass Sie zurechtkommen?
You're getting on top of this, aren't you?	Sie kommen damit klar, nicht wahr?
Everything's more or less okay with you, isn't it?	Es ist alles mehr oder weniger in Ordnung mit Ihnen, nicht wahr?
Mentioning good memories	**Gute Erinnerungen benennen**
You have a lot of good memories.	Sie haben eine Menge schöner Erinnerungen.
I'm sure there were lots of good memories.	Ich bin sicher, da gab es eine ungeheuerliche Menge schöner Erinnerungen.
There must be so many pleasant memories.	Da muss es viele schöne Erinnerungen geben.
She/he'll always be in our thoughts.	Sie/er wird uns immer in Erinnerung sein.

16.3.2 Beispieldialog: Dialogue concerning bereavement

Background: Conversation between the head of human resources Sam Bright and the employee Peter Hansen who has just lost his wife.

Sam Bright: Peter, I'd like to express my condolences. I'm sure the whole department feels with you on your loss.

Peter Hansen: Thank you.

Sam Bright: Let me also offer my support and assistance during this difficult time. Perhaps we could sit down briefly to arrange a few things?

Peter Hansen: Sure, I'd appreciate that.

Sam Bright: Okay. Please have a seat. Can I offer you anything – coffee, tea, water?

Peter Hansen: Coffee would be good. Thank you.

Sam Bright: First, let me ask you if you're managing?

Peter Hansen: Actually, I'm ok – under the circumstances. It is still a dreadful loss of course. I mean we were married for over 25 years …

Sam Bright: I'm sure you have a lot of wonderful memories.

Peter Hansen: Oh, we certainly have. And I think I'm already beginning to focus on that now. During the last couple of months the effects of her illness were very much in the foreground, of course.

Sam Bright: It's a difficult time. Coming to grips, mourning, adjusting … As I said, if there is anything I can help you with. Do you need some time away? Are there any jobs we could pass on to someone else?

Peter Hansen: Well. I think there is, yes. I'd like to have a week off, if that's possible. Part of my annual holiday.

Sam Bright: Actually, I don't think it's necessary to deduct your holiday time for this. You have worked a lot of overtime. There is still enough left for a week off. When would you like to go? Right away, I guess?

16.3.3 Kondolenzschreiben verfassen

Die folgenden Punkte sollten Sie beim Verfassen von Kondolenzbriefen beachten.

- The letter of condolence should be personal but also formal.
- It could include common memories.
- The letter should be hand-written on high quality stationery, if possible.
- Keep it short and simple.
- Write at the beginning your reaction to the death of the individual.
- Describe common memories and experiences in very positive words.
- Conclude the letter carefully by expressing sympathy with a phrase like »with our deepest condolences«.

16.3.4 Musterschreiben: Letter of condolence

Dear John,

Hearing about your terrible loss this morning, I wanted to send you my sympathy and condolences right away.

I am truly sorry that you have to experience such a difficult time as this. I also want to send condolences and special wishes from everybody in the team. They are as shocked as I am and they have told me to express their support to you.

Of course, at a time like this it is good to have a large family to turn to for help and assistance. If there is anything that we can do, we are most willing. Please ask if there's anything.

Let me also say that you should not worry about any of the tasks here. I have assigned Judy as your stand-in. She will take care of business during your absence.

Please take off the next three weeks to arrange things.

Best wishes,

[Signature]

16.4 Vokabeln – Vocabulary input

Nachfolgend finden Sie eine Auflistung wichtiger Fachbegriffe aus diesem Modul.

English	German
absence *I signed the contract in the absence of the department leader.*	Abwesenheit *Ich unterschrieb den Vertrag in Abwesenheit des Abteilungsleiters.*
audience	Publikum, Zuhörer
behalf; on ~ of	im Namen von
bereavement	Trauerfall
celebrate	feiern
character trait	Charaktereigenschaft
charge with *Jim was charged with the job of streamlining processes.*	beauftragen *Jim wurde mit der Aufgabe beauftragt, Prozesse zu straffen.*
chronicles	Chronik, Aufzeichnungen
circumstance	Umstand
compassion	Mitgefühl
compliment *I wish to compliment the entire workforce for their commitment.*	Lob, Anerkennung *Ich möchte der gesamten Belegschaft für ihr Engagement ein Lob aussprechen.*
condolence	Beileid
deepest sympathy *Can I offer you my deepest sympathy?*	tiefstes Mitgefühl *Darf ich Ihnen mein tiefstes Mitgefühl aussprechen?*
desperate	verzweifelt
drawn-out *These management meetings are always dreadfully drawn-out.*	langgezogen *Diese Managementbesprechungen sind immer schrecklich in die Länge gezogen.*
evil	Übel
gathering	Zusammenkunft
glory *He's seeking for glory. Nothing else.*	Herrlichkeit; Ruhm *Er sucht nach Ruhm. Sonst nichts.*
greet	grüßen
high-flying speech	hochtrabende Rede
honorary guest	Ehrengast

English	German
lost for words *I didn't know what to say. I was lost for words.*	nicht die richtigen Worte finden *Ich wusste nicht, was ich sagen sollte. Mir fehlten die Worte.*
memory	Erinnerung
mind; have in ~ *What exactly do you have in mind?*	im Sinn haben *Was genau haben Sie im Sinn?*
mutual	gemeinsam, beiderseitig
occasion	Anlass
outlook	Ausblick
permit *If you permit me, I'll show you the findings.*	erlauben *Wenn Sie erlauben, zeige ich Ihnen die Untersuchungsergebnisse.*
privilege	Privileg
prosperity	Erfolg, Wohlstand
push someone	jemanden unter Druck setzen
recharge one's batteries	seine Batterien wieder aufladen
record *We recorded all events.*	festhalten, protokollieren *Wir haben alle Ereignisse festgehalten.*
review	Rückblick
rostrum	Rednerpult
rough times *We are going through rough times.*	harte Zeiten *Wir erleben harte Zeiten.*
shared values	gemeinsame Werte
speaker	Redner
speech	Rede
stress *Let me stress the fact that I was never too late.*	betonen *Lassen Sie mich die Tatsache betonen, dass ich nie zu spät war.*
task	Aufgabe
track; stay on ~	auf dem richtigen Kurs bleiben
volunteer *Who's going to volunteer for this job?*	etwas ehrenamtlich tun, freiwillig tun *Wer meldet sich für diesen Job freiwillig?*
years of service	Dienstjahre

16.5 Übungen – Exercise time

Task 1: Create your own toasts.
Fit in these verbs and nouns to complete the toasts.

Verbs: propose, raise, celebrate, suggest, toast, rise

Nouns: agreement, future, toast, conclusion

1. Let's ________________ our glasses to the ________________ of our company.
2. I ________________ to ________________ the guests of our company.
3. I'd like to ________________ a ________________ to the ________________ of our contract.
4. I'd like to ________________ a toast to the company.
5. Let's ________________ our ________________ with a glass of champagne.

Task 2: Polite noises for speakers. Translate the German expressions.

1. I can say in all ________________ (*Offenheit*) that I am very pleased to have the ________________(*Ehre)* of proposing a toast to the company.
2. But the fact that makes it so easy and indeed, ________________ (*freudig*) a ________________(*Aufgabe)* is simply that it is out of office hours.
3. I'm not going to ________________ (*so tun als ob*) that I'm one of those ________________(*selten)* birds who live for their work.
4. I ask you to ________________ (*sich anschließen)* me in drinking to the company's continued ________________ *(Wohlstand)*.

Task 3: Working with prepositions. Put in the prepositions.

1. ________________ this occasion ...
2. Looking back ________________ the past years ...
3. Burn the candle ________________ both ends.
4. I can say ________________ all sincerity that ...
5. It was thanks ________________ your input.
6. It depends ________________ the company.
7. It is done ________________ a spirit of mutual cooperation.
8. ________________ behalf of the staff, I ...

ARBEITSMITTEL AUF MYBOOK.HAUFE.DE

Auf der Internetseite zum Buch finden Sie weitere Übungen zu diesem Modul sowie eine Muster-Lobrede und eine Tischrede.

17 Die Abmahnung

Zusammenfassung

Jemandem die gelbe Karte zu zeigen und abzumahnen, ist schon auf Deutsch nicht so einfach. Wie Sie dies auf Englisch tun und was Sie dabei beachten müssen, erfahren Sie in diesem Modul.

- Eine *Übersicht* zeigt Ihnen, wie Sie bei der Abmahnung von Arbeitskräften vorgehen sollten.
- Der anschließende *Gesprächsleitfaden* führt Sie sicher durch ein erstes informelles Mitarbeitergespräch, das Sie führen, um auf fehlerhaftes Verhalten bzw. eine Verletzung des Arbeitsvertrags hinzuweisen.
- Um die formelle Verwarnung geht es in dem *Leitfaden*. Sie finden Musterformulierungen für das Mitarbeiterespräch.
- Schließlich geben wir Ihnen *Textbausteine* für eine schriftliche Abmahnung von Mitarbeitern:innen an die Hand.

Am Ende des Moduls finden Sie alle wichtigen *Vokabeln* zu diesem Thema und *Übungsaufgaben*, mit denen Sie Ihre Kenntnisse überprüfen können.

Die Meilensteine: Korrekt abmahnen

Abb. 17: Korrekt abmahnen

17.1 Abmahnungsgespräche führen

Employees should follow a set of rules and regulations at the workplace. When a rule or law is broken, disciplinary action should be taken. The important word in the process is »fairness«.

Mitarbeitende sollten eine Reihe von Regeln und Vorschriften am Arbeitsplatz befolgen. Wenn eine Regel oder ein Gesetz gebrochen wird, zieht das Disziplinarmaßnahmen nach sich. Das wichtige Wort in diesem Prozess ist Gerechtigkeit.

The process contains rules and regulations, penalties or disciplinary measures, and the employee's right to appeal.

Der Prozess beinhaltet Regeln und Vorschriften, Strafen oder Disziplinarmaßnahmen, aber auch das Recht des Angestellten, dagegen vorzugehen.

17.1.1 Übersicht: Wie Sie bei einer Abmahnung vorgehen

1.	Manager has proof of an employee's bad behaviour or poor work.
2.	Have an informal talk to make sure he or she knows there has been a »violation«. Talk on how to solve it (apology, training).
3.	Monitor the employee's further activities.
4.	Some time later, has the situation got better?
5.	If no: Inform employee in writing that he/she must attend a formal interview giving reasons.
6.	Formal interview: What has been done to date? Employee explains his situation. Okay, not okay?
7.	If no: Inform employee that he will get a formal warning where it is written that the job may be at risk if the situation does not improve.
8.	Situation is improved: Send him/her your comments in writing. Warnings will be disregarded. If situation is not improved: No explanation given.
9.	If situation is not improved: Inform employee in writing of the 2nd formal interview, write a second warning.

17.1.2 Gesprächsleitfaden: Informelles Gespräch wegen schlechter Leistung

Dieser Leitfaden führt Sie durch ein informelles Gespräch mit Mitarbeitern:innen, um diese auf Probleme oder Pflichtverletzungen aufmerksam zu machen und ihnen die Möglichkeit zu geben, dazu Stellung zu nehmen. Am Anfang des Leitfadens finden Sie ein paar Tipps, die Sie in Mitarbeiteresprächen unbedingt beachten sollten.

Tips on dealing with discipline	Hinweise für den Umgang mit Kritik
Speak clearly and honestly.	Sprechen Sie deutlich und einfach.
Behave firmly but fairly.	Treten Sie fest, aber gerecht auf.
Make constructive suggestions.	Machen Sie konstruktive Vorschläge.
Always keep a record of your discussions.	Führen Sie immer Aufzeichnungen über Ihre Gespräche.
Stay calm at all times.	Bleiben Sie zu jeder Zeit ruhig.
Don't procrastinate – take necessary action.	Schieben Sie nichts auf – ergreifen Sie die notwendigen Maßnahmen.
Don't allow the issue to escalate.	Verhindern Sie eine Eskalation des Themas.
Explaining what the problem is	**Das Problem erläutern**
It's come to my notice that you've been reacting very slowly to customers' needs.	Mir ist aufgefallen, dass Sie sehr langsam auf Kundenanforderungen reagiert haben.
We've been hearing that you are finding it difficult to deal with customers.	Uns ist zu Ohren gekommen, dass Sie Schwierigkeiten im Umgang mit Kunden hatten.
It seems you are too unfriendly to customers.	Es sieht so aus, als wären Sie zu unfreundlich zu Kunden.
Look! The following things are maybe not acceptable. Firstly you …, then you …	Schauen Sie! Die folgenden Dinge sind möglicherweise nicht akzeptabel. Erstens haben Sie … dann haben Sie …
Softening the blow	**Die Kritik abschwächen**
Actually, I'm not really here to blame you or point a finger.	Eigentlich bin ich nicht hier, um Sie zu beschuldigen.
In fact, it's not a question of who is right or wrong.	Es geht nicht um die Frage, wer recht hat oder nicht.
There are one or two things we should talk about but it's nothing we can't solve.	Da gibt es ein oder zwei Dinge, über die wir reden sollten, aber nichts, was wir nicht lösen können.
Today we have to sort out what's true and maybe not true.	Heute müssen wir herausfinden, was wahr ist und vielleicht nicht wahr ist.
Suggesting solutions	**Lösungen vorschlagen**
Perhaps we should take a look at a specific example.	Vielleicht könnten wir uns ein einzelnes Beispiel anschauen.
Let's see if we can get a handle on all this.	Lassen Sie uns schauen, ob wir das geregelt bekommen.
Now we should think about the following.	Nun sollten wir über das Folgende nachdenken.
In my opinion, perhaps we should do the following.	Meiner Meinung nach, sollten wir vielleicht das Folgende tun.

Introducing examples	**Beispiele anführen**
Well, last week we had a customer from XX. We could see you kept him waiting.	Nun, letzte Woche hatten wir einen Kunden von XX. Wir konnten sehen, dass Sie ihn haben warten lassen.
To give you an idea we had the following situation: …	Um Ihnen eine Vorstellung zu geben, gab es die folgende Situation …
I can show you what I mean. We …	Ich kann Ihnen zeigen, was ich meine. Wir …
If I might explain, it was like this …	Wenn ich erklären dürfte, es war wie folgt …
A few words of discipline	**Ein paar tadelnde Worte**
Xxx, it's not our place to comment on the way our customers dress.	Es steht uns nicht an, die Art und Weise zu kommentieren, wie sich unsere Kunden kleiden.
Working in this way isn't productive. It causes problems.	Die Art und Weise, wie Sie arbeiten, ist nicht produktiv. Das verursacht Probleme.
We have to do our job, not play the role of a judge.	Wir müssen unsere Aufgaben erledigen, nicht die Rolle eines Richters übernehmen.
If I could just say something. Well, I'm afraid this has to stop.	Wenn ich etwas sagen könnte. Nun, es tut mir leid, aber das muss aufhören.
Softening the situation and suggesting solutions/making alternative suggestions	**Die Situation abschwächen und Lösungen vorschlagen/Alternativvorschläge unterbreiten**
You must understand that we cannot tolerate such behaviour.	Sie müssen verstehen, dass wir ein solches Verhalten nicht tolerieren können.
In this situation I'd suggest some training.	In dieser Situation würde ich eine Trainingsmaßnahme vorschlagen.
You realise that you have to know the customer.	Es ist ihnen klar, dass Sie die Kunden kennen müssen.
Perhaps we should consider planning everything completely differently.	Vielleicht sollten wir daran denken, alles komplett anders zu planen.

17.1.3 Beispieldialog: Informal talk with the employee about poor work

Background: The general office department supervisor, Cliff Oak, is informally warning the receptionist Sandra Norman about her work.

Cliff Oak: It's come to my notice that you have been reacting very slowly to customers' needs. Keeping them waiting. Treating them in an unfriendly, off-hand way. And you've also done this with colleagues.

Sandra Norman: Well, that's complete nonsense.

Cliff Oak: I know this is a difficult situation. But I'm not really here to blame you or point a finger. I just want to make sure that you understand that things aren't quite right. We have to make some changes to get back on track.

Sandra Norman: Well, I'm very surprised about all this.

Cliff Oak: Perhaps we could take a look at a specific example. You realise that this desk is monitored by CVT cameras. That's also part of the training.

Sandra Norman: Yes, of course I know that.

Cliff Oak: Well, last week we had a customer from Ally Life Insurances. We could see that you kept him waiting, maybe three or four minutes.

Sandra Norman: Well, I was dealing with some emails. And, by the way, he looked very scruffy.

Cliff Oak: Ms Norman, it's not our place to comment on the way our customers dress. In any case, the situation must improve. It can't go on.

Sandra Norman: I really didn't notice.

Cliff Oak: Well, your job is to notice. So let's see what we can do. I'd suggest some sort of training. We have a 3-day customer service seminar coming up. I'd like you to participate in that.

17.1.4 Musterschreiben: Letter inviting to attend a formal disciplinary interview

Dear Ms White,

During the period May 5 – May 20 we received various expressions of dissatisfaction with regard to your performance as a receptionist in the Research & Development building.

In order to support you in changing this situation the company conducted an informal discussion informing you about the situation and suggesting improvements, including a special training course, which you completed two weeks later.

In the six weeks following we monitored your activities according to the standard practices of this company.

However, I regret to inform you that your performance and behaviour have not improved to a satisfactory level. We are therefore obliged to invite you to a first »formal interview« to clarify this issue.

Best regards

[Signature]

17.1.5 Gesprächsleitfaden: Offizielles Abmahnungsgespräch

Hat ein informelles Mitarbeitergespräch keine Früchte getragen, folgt im nächsten Schritt ein offizielles Gespräch mit der Androhung disziplinarischer Maßnahmen. Nützliche Musterformulierungen für solche Gespräche finden Sie im folgenden Leitfaden.

Briefly introducing the topic	**Kurze Einleitung zum Thema**
Well, this is the second time that we've come together to ...	Nun, das ist das zweite Mal, dass wir zusammenkommen, um ...
We're together again to discuss the issue of ...	Wir sind jetzt wieder zusammengekommen, um über das Problem bezüglich ... zu sprechen.
To get straight to the point, you ...	Um gleich auf den Punkt zu kommen, Sie ...
I won't beat around the bush, you ...	Um es ohne viele Worte zu sagen, Sie ...
Summarising the issue	**Zusammenfassung der Angelegenheit**
In our first meeting I explained the problem. Your activities were not appreciated.	Bei unserem ersten Treffen habe ich das Problem erläutert. Das, was Sie getan haben, fand keine positive Resonanz.
I gave you some instructions which have not been followed.	Ich gab Ihnen einige Anweisungen, die Sie nicht befolgt haben.
To sum it all up, you were warned and have not reacted.	Um es zusammenzufassen, Sie wurden gewarnt und haben nicht reagiert.
To put it in a nutshell, you have not conformed to the rules.	Um es kurz zu sagen, Sie haben sich nicht an die Regeln gehalten.

Asking employees to show their improvement	**Mitarbeiter:innen bitten, Entwicklungsfortschritte aufzuzeigen**
Can you describe your improvements?	Können Sie mir Ihre Fortschritte beschreiben?
Could you explain how you see that things are better?	Könnten Sie mir erklären, inwiefern Sie jetzt eine Verbesserung sehen?
In your opinion, what's been better in the last XX months?	Was ist aus Ihrer Sicht in den letzten XX Monaten besser geworden?
I'd like you to show me where you see an improvement.	Ich hätte gerne, dass Sie mir zeigen, wo Sie eine Verbesserung sehen.
Final comment of the supervisor	**Abschließende Bemerkungen durch Vorgesetzte**
Well, I feel the whole situation isn't okay.	Ich denke, dass die ganze Situation nicht in Ordnung ist.
In total I'd say we've got a problem, wouldn't you?	Insgesamt würde ich sagen, dass wir ein Problem haben, sehen Sie das auch so?
As far as I can see this issue hasn't been solved.	So weit ich das sehen kann, besteht das Problem weiterhin.
To draw a line under this, I feel …	Um es zusammenzufassen, ich denke …
From my point of view I'm dissatisfied with what I've seen.	Von meinem Standpunkt aus bin ich mit dem, was ich gesehen habe, unzufrieden
Well, I regret to inform you that …	Nun, es tut mir leid Ihnen zu sagen, dass …

17.1.6 Beispieldialog: Formal disciplinary interview

Background: Interview between a member of the personnel department, Jane Eastmen, and Beverley White who is working as a receptionist in the company. Ms White has been formally warned already.

Jane Eastmen: Ms White. Thank you for coming. Well, this is the second time that we've come together.

Beverley White: Yes, that's right, and I'm still not quite sure why.

Jane Eastmen: Ms White, in our first meeting I explained the position of the company. That your activities were not appreciated and deemed not to fulfil the tasks which were expected of you.

Beverley White: Everybody has his or her own style of work. I spread myself between tasks.

Jane Eastmen: I'm sure that's how you see it, but we have a different view.

Can you describe your improvement since the training course you were on?

Beverley White: Well, I try to listen to what the people who come to Reception say. But I must say I didn't feel the training was any good.

Jane Eastmen: Oh, why do you say that?

Beverley White: Well, for example the trainer tried to do role-plays with me.

Jane Eastmen: That's a standard way of training behaviour issues, you see what your partner feels. That's part of learning.

Beverley White: But he gave me the role of the customer. That's not my job.

Jane Eastmen: Okay. So you feel there was no improvement.

Beverley White: Not really. The visitors are still all behaving the same way. As if I'm their slave.

Jane Eastmen: I see. Well, I feel the whole situation is not okay. I regret to inform you that you will be receiving a letter from me. And it will be a formal warning. If your performance does not come up to what is expected by the company, your job may be at risk.

17.2 Schriftlich abmahnen

The erring employee has attended a first formal interview. He or she was clearly and fairly informed that the company was dissatisfied with his or her work or behaviour and the job may be at risk.

Der:Die sich vertragswidrig verhaltende Mitarbeiter:in hat an dem ersten formellen Gespräch teilgenommen. Hier wurde er oder sie eindeutig und fair darüber informiert, dass die Firma mit seiner oder ihrer Arbeit beziehungsweise dem Verhalten nicht zufrieden ist und der Arbeitsplatz gefährdet ist.

Now follows a period of monitoring. If the required improvement is made, the employee will be sent a comment in writing which may include a statement that the previous warning will be disregarded. If the employee's activities are still not satisfactory, a second formal warning can be sent inviting the employee to a second formal interview.	Nun folgt eine Zeit der Beobachtung. Wenn die geforderte Verbesserung sich einstellt, wird der:die Mitarbeiter:in schriftlich darüber informiert, dass die zuvor erfolgte Warnung gegenstandslos geworden ist. Sollte sich nach wie vor immer noch keine Verbesserung eingestellt haben, ist es möglich, den:die Mitarbeiter:in eine zweite Verwarnung, die eine Einladung zu einem zweiten formalen Gespräch beinhaltet, zu schicken.

17.2.1 Textbausteine: Eindeutige Abmahnungsschreiben verfassen

Bleiben sämtliche Gespräche ohne Erfolg, so folgt in einem nächsten Schritt die schriftliche Abmahnung. Welche Inhalte die Abmahnung enthält sowie passende Musterformulierungen, zeigt der folgende Leitfaden.

Complaint: Defining the actual facts of misconduct or violation of the employment contract including the date and time	**Beschwerde: Den Sachverhalt des Fehlverhaltens bzw. des Verstoßes gegen den Anstellungsvertrag definieren unter Angabe des Datums und der Zeit**
You were asked to pay immediate attention to clients.	Sie wurden darum gebeten, sich unverzüglich um die Kunden zu kümmern.
On 1.5.2010 and 15.5.2010 you were found in our warehouse drinking alcohol. This is not allowed.	Am 1.5.2010 und am 15.5.2010 wurden Sie in unserem Lager angetroffen, als Sie Alkohol konsumierten. Dies ist nicht erlaubt.
A complaint has been filed regarding your repeated failure to attend department meetings.	Es wurde eine Beschwerde bezüglich Ihres wiederholten Fernbleibens von Abteilungsbesprechungen eingereicht.
Notification: Stating that repetition of the misconduct or violation will not be tolerated	**Mitteilung: Darlegen, dass die Wiederholung des Fehlverhaltens oder Verstoßes nicht toleriert wird**
If you continue with this misconduct we will be forced to terminate your contract.	Wenn Sie dieses Fehlverhalten fortsetzen, sind wir gezwungen, Ihren Arbeitsvertrag zu kündigen.
We cannot tolerate this sort of behaviour. This letter serves as a formal warning.	Diese Verhaltensweise wird von uns nicht hingenommen. Dieses Schreiben fungiert als formale Warnung.
We would like to make it clear to you that this sort of misconduct is not acceptable.	Wir möchten Ihnen deutlich machen, dass diese Art von Fehlverhalten nicht akzeptabel ist.

Informing employees that the letter is added to personnel file	Mitarbeiter:innen informieren, dass der Brief in die Personalakte aufgenommen wird
This formal warning will be included in your personnel file.	Diese formale Warnung wird Ihrer Personalakte beigefügt.
This letter will be added to your personnel records.	Dieses Schreiben kommt zu Ihren Personalakten.
This is to inform you that the warning will be recorded in your personnel file.	Hiermit informieren wir Sie, dass die Verwarnung in Ihrer Personalakte aufgenommen wird.
Announcement: Making aware that in case of repetition, dismissal will follow	**Ankündigung: Bewusst machen, dass im Wiederholungsfall die Kündigung folgt**
If this continues, the result will be dismissal.	Sollte sich dies fortsetzen, wird dies in einer Kündigung enden.
Your contract will be terminated if this is repeated.	Der Arbeitsvertrag wird gekündigt werden, sollte sich dies wiederholen.
Ms XX, please understand that if this goes on you will be dismissed.	Verstehen Sie bitte, Frau XX, wenn das so weitergeht, werden sie entlassen.
Further incidents of this nature will result in dismissal.	Weitere Vorfälle dieser Art werden zur Entlassung führen.

17.2.2 Musterschreiben: Written warning

Musterschreiben 1: Failed Performance

Please take note that this letter serves as a first reminder and a warning about your recent performance which unfortunately has failed to fulfil the expectations of the company.

Several complaints have been filed in the last 2 months on the late submissions of your department's monthly report. Efforts to gain your co-operation have proved unsuccessful. You also failed to attend when summoned to an emergency meeting on Tuesday, 10 May 2010 without a satisfactory explanation.

We would like to make it clear to you that we take the employee's performance very seriously and any indication of an unsatisfactory performance will be subject to close examination and, when necessary, disciplinary actions. Therefore, it would be very much appreciated if you would see this warning letter as a reminder and attempt to improve your performance.

Should you have any issues that you need to speak about, please do not hesitate to come forward and talk to me.

Yours sincerely,

[Signature]

Musterschreiben 2: Violation of department's work rules
This letter serves as a formal written warning.

This action is being taken as a result of your violation of this department's work rules regarding our procedures for dealing with suppliers and in particular the acceptance of gifts.

On 23 June 2009 you were also counselled regarding the violation of this work rule.

From this date forward, you are expected to observe our rules and procedures in purchasing.

Further incidents of this nature will result in more severe disciplinary action being taken, up to and including discharge.

[Signature]

17.3 Vokabeln – Vocabulary input

Nachfolgend finden Sie eine Auflistung wichtiger Fachbegriffe zu diesem Modul.

English	German
admonish *He was admonished for poor work.*	abmahnen; ermahnen *Er wurde für seine schlechte Arbeit ermahnt.*
appeal	anfechten
behave *You have behaved in accordance with company regulations.*	verhalten, sich benehmen *Sie haben sich gemäß den Unternehmensregeln verhalten.*
come straight to the point	gleich auf den Punkt kommen
complaint *I would like to make a complaint.*	Beschwerde *Ich möchte eine Beschwerde vorbringen.*

English	German
constructive suggestion	konstruktiver Vorschlag
counsel *The firm has several members of staff who can counsel you.*	beraten *In der Firma gibt es mehrere Mitarbeiter, die Sie beraten können.*
deem	etwas befinden; erachten als
discharge	Entlassung
dismissal *Drinking alcohol at your workplace will result in dismissal.*	Entlassung *Das Trinken von Alkohol am Arbeitsplatz führt zur Entlassung.*
disregard *She has continually disregarded the rules.*	ignorieren, missachten *Sie hat ständig die Regeln missachtet.*
draw a line under something *It was a bad year. Now it's time to draw a line under it.*	einen Schlussstrich unter etwas ziehen *Es war ein schlechtes Jahr. Nun ist es an der Zeit, darunter einen Schlussstrich zu ziehen.*
erring employee	Mitarbeiter auf Abwegen
file a complaint	eine Beschwerde einreichen
force	zwingen
handle; get a ~ on something	etwas in den Griff bekommen
immediate *He was dismissed with immediate effect.*	sofort *Ihm wurde fristlos gekündigt.*
improvement	Verbesserung
misconduct *There are strict rules in place to prevent misconduct by the employees.*	Fehlverhalten *Es gibt strikte Regeln, die ein Fehlverhalten der Mitarbeitenden verhindern sollen.*
monitor *We will monitor your behaviour closely over the next 6 months.*	überwachen *Wir werden Ihr Verhalten in den nächsten 6 Monaten genauestens überwachen.*
nutshell; to put it in a ~	kurz gesagt
obliged	verpflichtet
off-hand *The receptionist was very off-hand and unfriendly.*	ungezogen; unverschämt *Die Empfangsperson war äußerst unverschämt und unfreundlich.*
penalty	Strafe
procrastinate	aufschieben

English	German
point a finger *We will point a finger at the marketing department for the drop in sales.*	beschuldigen *Wir beschuldigen die Marketingabteilung für die sinkenden Verkaufszahlen.*
regulation	Anordnung, Bestimmung
reminder	Erinnerung
scruffy	ungepflegt
sensible way	vernünftige Weise
set of rules	Regelwerk
spread	verteilen
suggestion	Vorschlag
sum up	zusammenfassen
summon *Jane was summoned to a meeting.*	aufrufen, bestellen *Jane wurde zu einem Gesprächstermin bestellt.*
violation *This was a clear violation of the regulations.*	Verstoß *Dies war ein eindeutiger Verstoß gegen die Regeln.*

17.4 Übungen – Exercise time

Task 1: Complete the sentences with the appropriate translation of the German word in brackets.

1. Discipline is needed at the (*Arbeitsplatz*).
2. This warning is made (*schriftlich*).
3. We're certainly not here to (*beschuldigen*) the employee.
4. They may be fired for incapability or bad (*Verhalten*).
5. She works as the (*Abteilungsleiterin*) of the General Office.

Task 2: Put in the correct form of the word in brackets.

1. You should describe exactly what you saw. (*clear*)
2. I'm afraid we see your results a little . (*different*)
3. The company is rather with your progress. (*satisfy*)
4. Until now we haven't been able to notice any . (*improve)*
5. Please don't . Use every minute efficiently. (*procrastination*)

Task 3: Vocabulary grid

Noun	Verb	Adjective	Adverb	Opposite adjective
	be fair			
	improve		XXX	
specification			XXX	
	differ			
		surprised		
			possibly	

ARBEITSMITTEL AUF MYBOOK.HAUFE.DE

Auf der Internetseite zum Buch stehen Ihnen weitere Übungen zu diesem Modul sowie weitere praktische Arbeitsmittel zum Thema Abmahnung zur Verfügung:

- Vorlage für ein positives Feedbackgespräch nach einer formellen Warnung.
- Vorlage für die erneute Einladung eines:r Mitarbeiters:in zu einem Disziplinargespräch.

18 Die Kündigung

Zusammenfassung

Arbeitskräften zu kündigen ist in den meisten Fällen nicht einfach. Wie Sie dabei vorgehen können, zeigen wir Ihnen in diesem Modul. Zunächst geben wir Ihnen einen Überblick für das schrittweise Vorgehen bis hin zur Entlassung.

- In einem *Gesprächsleitfaden* mit *Beispieldialog* können Sie lesen, wie Sie ein professionelles Kündigungsgespräch führen.
- Wir geben Ihnen *Textbausteine* und ein *Musterschreiben* für die schriftliche Kündigung an die Hand.
- Und Sie finden im zweiten Teil des Moduls einen *Gesprächsleitfaden,* falls Sie die Kündigung direkt am Arbeitsplatz vollziehen müssen.

Am Ende des Moduls finden Sie alle wichtigen *Vokabeln* zu diesem Thema und *Übungsaufgaben*, mit denen Sie Ihre Kenntnisse überprüfen können.

Die Meilensteine: Wie Sie bei einer Kündigung vorgehen

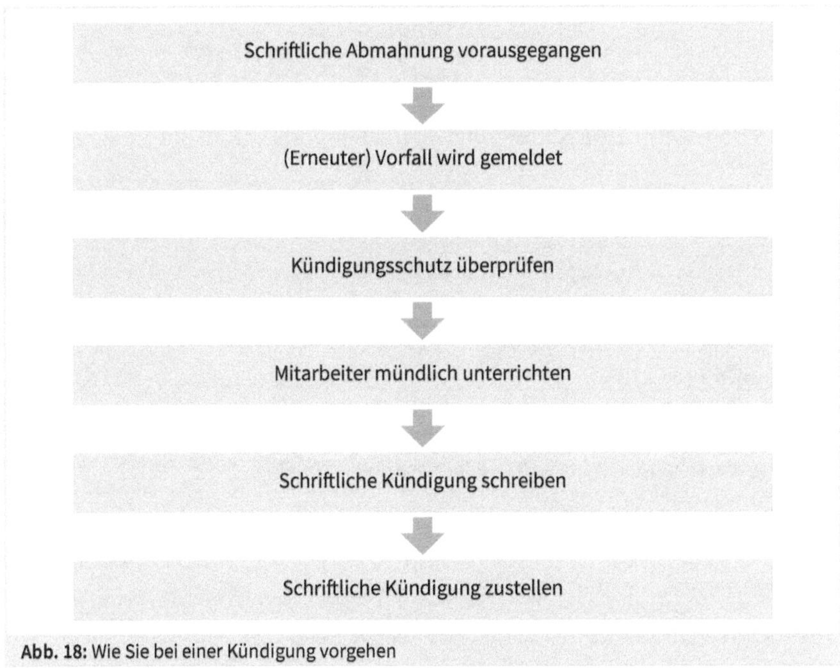

Abb. 18: Wie Sie bei einer Kündigung vorgehen

18.1 Kündigung wegen Minderleistung

Typische Kündigungssituationen

Dismissal takes place in the following situations: The contract of employment is terminated by the employer either with or without notice. The employee terminates the employment contract with or without notice. This is due to something the employer has done. This action is called »constructive dismissal«!	In folgenden Situationen kommt es zur Entlassung: Der Arbeitsvertrag ist durch den Arbeitgeber entweder fristgemäß oder fristlos gekündigt worden. Der:Die Arbeitnehmer:in kündigt den Arbeitsvertrag fristgemäß oder fristlos. Dies hängt mit dem Verhalten des Arbeitgebers zusammen. Dieses Verhalten wird als »konstruktive Kündigung« bezeichnet.
The employee has a fixed-term contract and the term expires. There is no renewal of the same contract.	Der:Die Arbeitnehmer:in hat einen zeitlich befristeten Arbeitsvertrag und die Frist ist ohne Erneuerung desselben Vertrags abgelaufen.

18.1.1 Übersicht: Eine Kündigung wegen Minderleistung durchführen

Diese Übersicht schließt direkt an die Übersicht aus Modul 17 »Die Abmahnung« an.

1.	*1st formal interview*: Repeat earlier steps for employee. Is employee's explanation acceptable?
2.	*If no*: Inform employee you will give him a formal warning informing him that his employment may be at risk.
3.	Continue to monitor employee.
4.	Has the employee improved his performance sufficiently?
5.	Is there alternative employment in the company?
6.	*If no*: Invite employee to a 2nd formal interview and give him a 2nd written warning. Follow statutory dismissal and disciplinary procedure.
7.	If the decision is within the range of reasonable responses of a reasonable employer.
8.	Dismiss the employee in writing with notice for pay in lieu of notice. Inform him he has the right to appeal.
9.	If he exercises his right to appeal it should be heard by a member of senior management.

18.1.2 Gesprächsleitfaden: Ein Kündigungsgespräch führen

Hier erhalten Sie Musterformulierungen, um Kündigungsgespräche zu führen.

Giving general advice	**Allgemeine Ratschläge geben**
I understand your issue. I have to explain that you need to keep the matter factual.	Ich verstehe Ihr Anliegen. Ich muss voraus-schicken, dass Sie die Angelegenheit sachlich sehen müssen.
I'm aware of your situation, but you'll have to keep to the facts.	Mir ist ihre Situation bewusst, aber Sie werden bei den Fakten bleiben müssen.
I'm informed about your problem, but please keep to the point, stay fair.	Ich bin über Ihr Problem informiert, aber blei-ben Sie bei der Sache, bleiben Sie fair.
I've been thinking about your dismissal matter. Firstly it's necessary to be fair, second to keep to the facts, third to act quickly.	Ich habe über Ihre Entlassung nachgedacht. Zuerst sollte man fair, dann sachlich bleiben und an dritter Stelle schnell reagieren.
Introducing the question of law	**Die Rechtslage ins Spiel bringen**
Well, I have to tell you straight away that according to the law you can't fire someone because ...	Nun ich muss Ihnen geradewegs sagen, dass Sie rechtlich gesehen niemanden aufgrund von ... kündigen können.
To start off, the law has rules about dismissal. You can't simply do it.	Zuerst einmal, es gibt gesetzliche Kündigungs-regeln. Sie können das einfach nicht machen.
I should inform you from the outset that there is a legal procedure.	Zuallererst sollte ich Ihnen sagen, dass hier ein rechtlich vorgeschriebenes Verfahren besteht.
The law says you have to follow a procedure if you want to terminate your contract.	Wenn Sie kündigen wollen, schreibt Ihnen das Gesetz ein Verfahren vor.
Describing what is legally involved	**Die Gesetzgebung erläutern**
Financial compensation, clear job description and any sackable offences have all been docu-mented in it.	Finanzielle Entschädigung, eindeutige Arbeits-platzbeschreibungen und jede Art von Ver-stößen, die zur Kündigung führen, sind darin festgelegt.
The law stipulates/says that you must include financial compensation, transparent job de-scription and sackable offences.	Das Gesetz verlangt, dass Sie finanzielle Entschädigung, transparente Arbeitsplatzbe-schreibungen und Verstöße, die zur Kündigung führen, einbeziehen müssen.
Describing sackable offences	**Entlassungswürdige Verstoße beschreiben**
If they regularly don't perform the tasks in the job description.	Wenn sie wiederholt den Anforderungen aus der Arbeitsplatzbeschreibung nicht gerecht werden.
When they regularly don't carry out the tasks stated in the job description.	Wenn sie wiederholt nicht die Aufgaben aus der Arbeitsplatzbeschreibung ausführen.

Constant late arrival can be a reason for dismissal.	Ständiges Zuspätkommen kann einen Kündigungsgrund darstellen.
Repeatedly poor quality work can be a cause for dismissal.	Wiederholte schlechte Arbeitsergebnisse kann ein Kündigungsgrund sein.
If the employee constantly argues with colleagues.	Wenn ein:e Mitarbeiter:in sich permanent mit Kollegen streitet.

18.1.3 Beispieldialog: Dismissal of employees

Background: This is a dialogue between the HR Consultant Jefferson Morgan and the Managing Director Susan Thompson of the service company about firing employees. Issue: The Managing Director Susan Thompson has to dismiss two members of staff because she considers they have not come up to »expectations«.

Jefferson Morgan: So, I understand your issue. I have to explain to you from the beginning that you, the employer, need to keep this matter factual, you need to be fair, and you should do it quickly but not in a rushed way.

Susan Thompson: I know what you mean, but how do I do it?

Jefferson Morgan: Well, I have to tell you straight away that legally you can't fire somebody because you think they haven't come up to »expectations«, as you put it. There is a transparent list of rules for this that you have agreed with the union, about dismissal and redundancy.

Susan Thompson: What do these rules contain?

Jefferson Morgan: They have set out financial compensation, a clear job description, any sackable offences, e.g. working for competitors.

Susan Thompson: What can I fire them for, then?

Jefferson Morgan: Well, if they regularly do not perform the tasks described in their job descriptions. If they constantly arrive late, do poor quality work, pay no attention to detail, or if they are always arguing with other employees.

Susan Thompson: Well, that's certainly the case.

Jefferson Morgan: It's not so simple. Those were some reasons but only provided that you've given the defaulters verbal, then written warnings about their work and behaviour, and informed them that they must improve or they will face dismissal.

18.1.4 Textbausteine: Kündigung schriftlich formulieren

Nach dem Gespräch mit dem Mitarbeiter, dem gekündigt werden soll, folgt das schriftliche Kündigungsschreiben. Diese Übersicht zeigt Ihnen Aufbau, Inhalt sowie nützliche Musterformulierungen.

Salutation	**Anrede**
Dear Ms Miller/Dear Mr Smith	Sehr geehrte Frau Miller/Sehr geehrter Herr Smith
Dear Dr Block	Sehr geehrter Herr Dr. Block
Referring to the context	**Bezugnahme auf den Inhalt**
I refer to our meeting on …	Ich beziehe mich auf unser Treffen vom …
With reference to our meeting of …	Unter Bezugnahme auf unser Treffen vom …
I'd like to refer to our meeting of …	Ich möchte mich auf unser Treffen vom … beziehen.
I'm contacting you with regard to/regarding/in regard to our meeting of …	Ich kontaktiere Sie unter Bezugnahme auf unser Treffen vom …
Recapitulating the problem's history	**Zusammenfassung des Problems**
As I explained at the meeting, you were unable to …	Wie ich in unserem Treffen erklärte, waren Sie nicht in der Lage …
As was discussed with you …	Wie mit Ihnen besprochen wurde, …
To recapitulate the situation up to now. You …	Um die Situation bis dato zu rekapitulieren. Sie …
To formally repeat what has taken place until now. On [Date] you were …	Um formal nochmals zu wiederholen, was bis jetzt passiert ist. Am [Datum] waren Sie …
Repeating the action taken	**Wiederholung der ergriffenen Maßnahmen**
As you know we provided you with …	Wie Sie wissen, haben wir Ihnen … gegeben.
As you are aware we offered you …	Wie Ihnen bewusst ist, haben wir Ihnen … angeboten.

We have done everything to find …	Wir haben alles getan, um … zu finden.
The company has tried to …	Das Unternehmen hat versucht zu …
Repeating the employee's entitlement	**Den Anspruch des Mitarbeiters wiederholen**
During your notice period you will receive …	Während ihrer Kündigungsfrist werden Sie … erhalten.
You are entitled to be paid in full including …	Sie haben Anspruch auf volle Bezahlung einschließlich …
Your entitlement includes the following: …	Der Anspruch umfasst Folgendes: …
You have the right to receive the following: …	Sie haben auf Folgendes ein Recht: …
Reminder about the appeal procedure	**Erinnerung über die Möglichkeit Einspruch zu erheben**
I would like to take this opportunity to …	Ich möchte die Gelegenheit wahrnehmen, um …
I'm taking this opportunity/chance to …	Ich nehme diese Gelegenheit wahr, um …
At this point I would like to remind you that you are entitled to appeal.	An dieser Stelle würde ich Sie gerne daran erinnern, dass es Ihnen freisteht, dagegen vorzugehen.
In this letter I'd like to remind you that you have a right to appeal.	In diesem Schreiben möchte ich Sie daran erinnern, dass Sie das Recht zum Widerspruch haben.
Stating regrets	**Bedauern zum Ausdruck bringen**
We regret that we have been obliged to take this action.	Wir bedauern, dass wir diesen Schritt haben gehen müssen.
It is with regret that we have had to take this action.	Wir bedauern, diesen Schritt gehen zu müssen.
The Company regrets this unfortunate situation.	Das Unternehmen bedauert diese unglückliche Situation.
It is with some sadness that we have been obliged to follow this procedure.	Mit einigem Bedauern waren wir gezwungen, diesem Verfahren zu folgen.

18.1.5 Musterschreiben: Dismissal regarding capability (poor performance)

Dear [Name],

I refer to our meeting on 10 September 2009.

As I explained at the meeting, you have been unable to perform your duties to the standards required by the Company. Therefore, we have no alternative but to terminate your employment with the Company with effect from [Date].

As you know, we have provided you with training and the help to enable you to perform better in your job but without success. Moreover, we have attempted to find suitable alternative employment within the Company. We regret to inform you that nothing is available.

You are entitled to be paid in full, including any accrued holiday pay, during your notice period.

I would like to take this opportunity of reminding you that you are entitled to appeal against this decision through the Company's Disciplinary Procedure. If you wish to exercise this right, you must inform me within two working days of receiving this letter.

We regret that we have been obliged to take this action. We should like to thank you for your past efforts for the Company and wish you every success for the future.

Yours sincerely

[Signature]

18.2 Eine Kündigung am Arbeitsplatz durchführen

When it finally comes to a dismissal, i. e. all the steps have been followed and completed, it is time for the termination interview. This is a very difficult task. The interview itself should be planned very carefully due to its sensitivity. The interviewer should make sure that the employee keeps the appointment. No critical information should be given over the phone.

Wenn es schließlich zu einer Entlassung kommt, d. h. alle Schritte eingehalten und abgeschlossen wurden, ist es Zeit für ein Kündigungsgespräch. Das ist eine sehr schwierige Aufgabe. Aufgrund der Sensibilität dieses Gesprächs muss es sehr sorgfältig im Voraus geplant werden. DerGesprächsführer sollte sicherstellen, dass der Gesprächstermin wahrgenommen wird. Kritische Informationen sollten nicht am Telefon übermittelt werden.

18.2.1 Checkliste: Kündigungsgespräch vorbereiten und durchführen

When preparing the termination interview you should take the following points into account	**Diese Punkte sollten Sie beachten, wenn Sie sich auf ein Kündigungsgespräch vorbereiten**	
Plan the interview carefully.	Planen Sie das Gespräch sorgfältig.	
Make the employee keep his appointment.	Stellen Sie sicher, dass der:die Mitarbeiter:in den Termin wahrnimmt.	
Don't give any information to the employee on the phone.	Geben Sie dem:der Mitarbeiter:in keinerlei Informationen am Telefon.	
Allow approximately 10 minutes for the interview.	Planen Sie ca. 10 Minuten für das Gespräch ein.	
Use a neutral location, not your own office.	Treffen Sie sich an einem neutralen Ort, nicht in Ihrem Büro.	
Prepare in advance – any agreements, HR file, release announcement.	Bereiten Sie sich vor – sämtliche Vereinbarungen, die Personalakten, Freistellungsankündigung.	
Be available after the interview in case there are questions or problems.	Planen sie zusätzliche Zeit ein, um in Anschluss an das Gespräch auf Fragen oder Probleme einzugehen.	

During the interview you should pay attention to the following	Während des Gesprächs gilt es, Folgendes zu beachten	
Come quickly to the point.	Kommen Sie direkt zur Sache.	
Describe the situation.	Schildern Sie die Situation.	
Suggest where he or she should go next.	Machen Sie einen Vorschlag, wo er oder sie im Anschluss des Gesprächs hingehen soll.	
Review the severance package.	Sprechen Sie über die Abfindung.	

18.2.2 Gesprächsleitfaden: Kündigung am Arbeitsplatz durchführen

Nützliche Redewendungen für ein Kündigungsgespräch erhalten Sie im nachfolgenden Leitfaden.

Describing the situation	Schildern der Situation
You've clearly been following what has happened.	Sie haben das, was bis jetzt passiert ist aufmerksam verfolgt.
So, we've been through the whole process.	Nun sind wir durch den gesamten Prozess durch.
We did ..., and ..., and it hasn't brought any positive results, so ...	Wie haben ... und ... gemacht und es hat nichts gebracht, also ...
I'm sorry to say, the whole situation hasn't been very good. So ...	Es tut mir leid zu sagen, dass die ganze Situation sich nicht gut darstellt. Daher ...
Listening to the employee	**Mitarbeiter:innen zuhören**
I agree with you, you ...	Ich stimme Ihnen zu, Sie ...
Well, you seem to be coping well with the situation.	Es sieht so aus, als ob Sie mit der Situation gut zurechtkommen.
You're more or less okay with what I've told you, aren't you?	Sie kommen mehr oder weniger mit dem klar, was ich Ihnen gesagt habe, nicht wahr?
You seem to be telling me that ...	Mir scheint, als wollen Sie mir sagen, dass ...

Suggesting what the employee should do next	Mitarbeiter:innen den nächsten Schritt vorschlagen
Well, perhaps you should go to …	Nun vielleicht wollen Sie zu … gehen
So, it might be better if you went to HR and …	Also, es ist vielleicht besser, wenn Sie zur Personalabteilung gehen und …
After all this stress why don't you go to the cafeteria and I'll accompany you, if you like.	Nach all dem Stress, warum gehen Sie nicht in die Cafeteria und ich begleite Sie, wenn Sie wollen.
Well, I know it's difficult. Can I accompany you to …	Nun, ich weiß es ist schwierig. Kann ich sie zu … begleiten
Reviewing the severance package	**Über die Abfindung sprechen**
The severance package includes financial compensation, ten days paid holiday, …	Das Abfindungspaket beinhaltet eine finanzielle Entschädigung, 10 Tage bezahlten Urlaub, …
The basic award is calculated according to the employee's age, length of continuous service and gross average weekly wage.	Die Grundentschädigung errechnet sich aufgrund des Alters des Mitarbeiters, Länge der ununterbrochenen Firmenzugehörigkeit und des durchschnittlichen wöchentlichen Bruttoverdienstes.
Her severance will be a calculation based on …	Die Berechnung ihrer Abfindung basiert auf …

18.2.3 Beispieldialog: Termination interview

Background: A discussion between HR supervisor Marc Davis and Joe Fancy who works as a purchasing clerk.

Marc Davis: Well, I'm glad we managed to keep this appointment on what is a very unfortunate occasion.

Joe Fancy: So, I guess you're still not satisfied with my work.

Marc Davis: Okay, let's not waste any time. I called you here precisely because we've seen no improvement, at least not enough for the company.

Joe Fancy: But I didn't really have enough time to make changes.

Marc Davis: Can we come straight to the point? We've been through the whole process already. We invited you to a first formal interview, you had the first warning. It didn't bring positive results.

Joe Fancy: Oh, I didn't realise it was all so serious. I was doing things my own way. That's all. My goodness!

Marc Davis: Are you more or less okay with what I've told you? It's unfortunate it's come to this. You were warned and everything took place in a due and timely manner. I have to inform you now formally that the company will terminate your contract of employment as of 15 November 2009. You will receive pay in lieu of notice but we believe in the interests of yourself, the company and your colleagues that you should leave immediately. You will receive any days of vacation due to you up to that date. We regret that it has come to this.

Joe Fancy: For goodness sake. I wasn't expecting this. I …

Marc Davis: I understand, but it all follows the procedure. I would like to remind you that you are entitled to appeal. You should do that through a senior executive.

Joe Fancy: Oh, okay. I understand. What should I …?

Marc Davis: My advice is to go back to your office, collect some of your things and take the rest of the day off.

18.3 Vokabeln – Vocabulary input

Nachfolgend finden Sie alle wichtigen Begriffe und Fachbegriffe aus diesem Modul.

English	German
accrue *How many days of holidays have you accrued?*	angefallen, aufgelaufen *Wie viele Urlaubstage sind aufgelaufen?*
appeal	Einspruch
appointment; keep an ~	einen Termin einhalten
average	durchschnittlich
basic award	Grundentschädigung
compensation; financial ~ *You'll receive financial compensation when you leave.*	Abfindung *Sie erhalten eine Abfindung, wenn Sie gehen.*
conduct	Handlungsweise, Verhalten
constructive dismissal	konstruktive Kündigung

English	German
continuous service *He leaves the company after 30 years of continuous service.*	ununterbrochene Firmenzugehörigkeit *Er verlässt das Unternehmen nach einer ununterbrochenen Firmenzugehörigkeit von 30 Jahren.*
contract of employment	Arbeitsvertrag
defaulter	Schuldige(r)
dismissal	Kündigung
due *The payment is due in 10 days.*	fällig *Die Zahlung ist in 10 Tagen fällig.*
due and timely manner	ordnungsgemäß und fristgerecht
duty	Aufgabe / Pflicht
effect; with ~ from	mit Wirkung vom
entitled	Anspruch haben
entitlement	Anspruch
expire	auslaufen, enden
face something *The company may have to face fierce competition.*	mit etwas rechnen müssen *Die Firma muss mit einem harten Konkurrenzkampf rechnen.*
factual	sachlich
goodness sake; for ~	um Himmels Willen
gross	brutto
holiday pay	Urlaubsgeld
immediately	sofort
notice of dismissal	Kündigung
notice period *The notice period for both parties is 3 months.*	Kündungsfrist *Die Kündigungsfrist beträgt für beide Parteien 3 Monate.*
notice; without ~	fristlos
offence	Verstoß
recapitulate	rekapitulieren, wiederholen
redundancy	betriebsbedingte Kündigung
renewal	Erneuerung

English	German
sackable *Stealing is a sackable offence.*	entlassungswürdig *Diebstahl ist ein entlassungswürdiges Vergehen.*
severance package	Abfindungspaket
statutory *You'll be a member of the statutory health insurance.*	gesetzlich *Sie werden ein Mitglied der gesetzlichen Krankenversicherung sein.*
termination interview	Kündigungsgespräch
timely manner	fristgerecht
wage *As a blue-collar worker you'll receive a wage.*	Lohn *Als eine gewerbliche Arbeitskraft erhalten Sie einen Lohn.*

18.4 Übungen – Exercise time

Task 1: Vocabulary expansion

Noun	Verb	Adjective	German
1.	terminate		
2.	renew		
3. advice			
4.		involved	
5.	expect		
6.		dismissed	
7.	proceed	XXX	
8.	receive		

Task 2: Which is the correct word?

1. All steps have been *observed/followed* and completed.
2. HR has been reviewing the severance *package/pack.*
3. The employee said he hadn't realised it was all so *seriously/serious.*
4. I would like to *remember/remind* you that there is an appeal process.
5. You are *right/entitled* to appeal.

Task 3: Find the correct prepositions.

1. It is due ________________ something the employer has done.
2. There is no renewal ________________ the contract.
3. I should inform you ________________ the outset there is a procedure.
4. Tell him quickly but not ________________ a hurried way.
5. Give them written warnings ________________ their work.
6. ________________ your notice period you will be entitled to your usual pay.
7. You can appeal ________________ this decision.
8. Thank you ________________ your past efforts.

ARBEITSMITTEL AUF MYBOOK.HAUFE.DE

Auf der Internetseite zum Buch finden Sie weitere Übungen zu diesem Modul sowie eine Mustervorlage zur betriebsbedingten Kündigung.

Anhang

Zusammenfassung
Im Anhang finden Sie die folgenden Arbeitshilfen:

- Small talk around the world – Die kleine Smalltalk-Fibel
- Lexikon der Berufs- und Abteilungsbezeichnungen
- Glossar gebräuchlicher Abkürzungen
- Praktische Phrasen und Redewendungen im Personalwesen
- Wörterbuch Englisch-Deutsch
- Lösungen zu den Übungen im Buch

Small talk around the world

Smalltalk ist für den Umgang mit Kollegen und Geschäftspartnern im internationalen Geschäft sehr wichtig. Smalltalk bietet Ihnen die Möglichkeit, die Menschen, mit denen Sie zu tun haben, aus einer anderen Perspektive kennenzulernen. Zu diesem Zweck haben wir eine kleine Smalltalk-Fibel zusammengestellt – mit Notizen zu vielen verschiedenen Ländern und Regionen. Selbstverständlich kann eine solche Darstellung in wenigen Zeilen der kulturellen Vielfalt der jeweiligen Länder niemals ganz gerecht werden. Das liegt in der Natur einer übersichtlichen Beschreibung. Seien Sie sich also dessen bewusst, dass diese Beschreibungen immer auch etwas Stereotypes haben. Sie werden merken, dass Sie sich von diesen Stereotypen um so leichter lösen können, je besser Sie die Menschen und die Kultur eines Landes kennen.

Pride and prejudice
Most people are proud of their country and think that it is the best. They are convinced that their way of life, their literature, their cuisine is superior to those of the rest of the world. It is not easy to understand another culture and it is sometimes painful to adapt to it.

Often we don't have enough time to discover and appreciate the ways of other nations. That is why we are critical of their food, manners and politics. Criticism without understanding leads to ethnic stereotypes. When we stereotype a nation and its people we

- put them into categories
- attach labels
- become subjective
- forget that we are dealing with individuals
- reinforce our own prejudices.

Stereotyping is a barrier to mutual understanding because
- barriers grow up subconsciously and affect our attitude,
- it creates mistrust and hostility,
- people have no chance to prove that stereotypes are wrong,
- prejudices spread quickly, they are contagious.

Discover the country and keep an open mind. Form an opinion of your own.
- Don't listen to friends who have had negative experiences in a foreign country. They might have been unable to adapt.
- Don't make comparisons with how things are done at home. It will only increase your frustration.
- Don't spend all your time in offices and hotels. Get out, make contacts and discover the country.
- Don't expect to be liked.

Read the following pages with a critical mind. We have mixed reality with irony. Discover the stereotypes.

Small talk in England

Georges Mikes gives a good example that shows the difference between small talk on the Continent and small talk in England. A continental gentleman seeing a nice picture may remark:

»This view reminds me of Utrecht, where the treaties were signed on the 11th April, 1713, after the War of Spanish Succession. The river there, however, recalls the Guadalquivir which rises in the Sierra de Cazorla and flows into the Atlantic Ocean and is 650 kilometres long. Oh, rivers ... What did Pascal say about them? Les rivières sont les chemins qui marchent.«

A Briton would remain silent for some time and think about how he could put his feelings into words and then say »It's pretty, isn't it?« In England it is bad manners to be clever. The educated Briton is modest and simple.

While the American businessperson is straightforward and says what he thinks, his British colleague is a good listener and keeps back his position as long as possible. As Heinrich Heine once said: »Silence can be defined as a conversation with an Englishman.« He rarely disagrees openly with your proposals. He might say to a salesman: »I quite like this little device« only to go on with, »however, I'm not sure that it is of any use to my family.«

Mind your manners

- Formal greetings and handshaking are customary.
- A reserved manner is expected.
- Business cards are exchanged as a matter of courtesy.
- Avoid personal questions. Neutral topics of opening conversation are more acceptable than exchanging personal information at the first meeting.
- Religion and politics are best avoided at the beginning.
- They don't use hand gestures to emphasise points and are critical of those who do.
- They won't complain when things go wrong; they will however moan and tell other people when they get home. So provide an opportunity for them to say how they feel and if anything is or was not to their liking.

Small talk in the USA

»They don't stand on ceremony. They make no distinction about a man's background, his parentage, his education. They say what they mean, and there is a vivid muscularity about the way they say it ... They are always the first to put their hands in their pockets. They press you to visit them in their own home the moment they meet you, and they are irrepressibly good-humoured, ambitious and brimming with self-confidence in any company. Apart from that I've got nothing against them.«

Tom Stoppard, Dirty Linen, 1976

Mind your manners

- They are quickly on first-name terms with you.
- Don't insist on titles.
- Small talk is important especially if it is about their trip to Europe.
- A friendly, firm handshake with good eye contact is important.
- Smiling into the eyes is very important.
- Use positive, animated helpful language. (»No problem! I'll do that right away!«)
- When they ask you »How do you like the USA?« they don't expect you to give a frank and honest answer. They are patriots. Don't mention the negative sides of the country (e.g. the situation of racial minorities, Vietnam war, Gulf war). Think positive!
- Avoid religion and politics.
- Avoid smoking.
- Don't get too close to them. Give them space.
- They like to do business while eating. Time is money.

Small talk in France

They don't small talk at meetings which is so typical of the Americans or the British. Neither is humour used to break the ice or to warm up the audience. So do not sprinkle your presentation with jokes, because business is a serious matter.

Mind your manners

- The French are proud of their language and prefer to speak it with guests from abroad. Even if your French is bad, it helps to break down barriers.
- The French are formal and may be seen as cold and aloof. Use Monsieur and Madame regularly. Stay warm and friendly without getting over-familiar.
- Avoid conversation openers that are popular in England, Germany and the States: »What do you do for a living?«, »Are you married?«, »Do you have children?«. Keep to the Tour de France, their excellent food or discuss art and culture, if you are a master of the subject. The French tend to butt in on each other's conversations. This is not rudeness, but proof that they are listening and interested in taking part.

French people believe that there is more to life than the job. Eating is part of the secret to good living. They honour guests at the best restaurants with three to six-course meals. An American who talks business to a French person over dinner will find his French colleague wants to enjoy his meal. They do not admire workaholics, so don't talk business over lunch.

Small talk in Portugal

If the Portuguese people were not very different from the Spanish, Portugal would not exist. The Portuguese are an Atlantic nation, the Spanish are a Mediterranean nation. Until the 1974 revolution Portugal looked towards Africa (Angola, Mozambique) and South Africa.

Contrary to the Mediterranean peoples they are good speakers of the English language. Meetings normally start with some small talk.

Mind your manners

- They are not as sensitive about race or religion as the Spanish. They are not as emotional as the Italians or the Spanish.
- They are more formal than the Spanish. They address anyone who appears qualified or intelligent as Duotor or Engenheiro.
- Don't ask questions about the revolution, don't praise their neighbour, Spain.
- Keep to the beauty of the country and their successful football team Benefica Lisboa.

- They avoid conflicts with colleagues and business partners.
- They are friendly, cheerful and communicative even at the beginning of a relationship.

Small talk in Spain

Demonstrate interest in the country's history and culture. If you don't care for bullfighting and it comes up in conversation, avoid criticising. The Spanish consider it an art to confront a massive animal that has been trained to be killed by a man holding a small sword and a red cape. The typical English fox hunt is seen as a cowardly sport: the ladies and gentlemen sit safely astride a hose while thirty dogs tear a small fox to pieces.

Mind your manners

- Shake hands when first meeting. Wait for a woman to offer her hand first.
- Don't try to do business during afternoon siesta. Business lunch starts at about 13.30 and can last until 17.00. Dinner is between 22.00 and 23.00. Be careful not to get drunk. Dress conservatively.
- If invited to the home, a gift is not expected, but you may bring a box of sweets. Do not send flowers except on special occasions.

One peninsula – two nations	
Spain	**Portugal**
The Spanish are a Mediterranean nation.	The Portuguese are an Atlantic nation.
They think the Iberian Peninsula is Spanish.	They see the Spanish as potential invaders.
They are less formal. They use the tu form all over the place.	They are more formal. They use titles.
Personal dignity is more important than time or money. Do not make them lose face.	National honour, race, religion are not major factors.
They are »machos«.	They are less robust.
They like physical and eye contact.	They are more reserved.

Small talk in Russia

Russians are searchers of their Slavic souls. That is why other people sometimes find their Russian partners sentimental and gloomy. Conversation is rarely trivial. Within minutes, the subject is the meaning of life.

Mind your manners

- Introductions are formal and direct. A smile is only used for greeting personal friends. Smiling is not respectful on formal occasions and smiling for no reason is a sign of idiocy.
- Shake hands when first meeting people and leaving, but never across a threshold. It is a sign of bad luck.
- Russians are still very title-conscious: use titles and show respect.
- Russians think of themselves as being absolutely honest and trustworthy. So don't show distrust.
- Like most nations they are proud of their country. Don't tell them that in your country everything is bigger, cleaner and better organised.
- If you are invited to a Russian home, it is customary to bring flowers, vodka or wine.
- Dinner is eaten early, about six o'clock or whenever they feel hungry.
- Toasts are common, and the guest must be prepared to return toasts.
- Among younger businessmen you will find more and more non-smokers, teetotallers and even vegetarians. Invitations could include a ballet, circus or concert and dinner in a restaurant.

Russians are puzzled that westerners can spend the equivalent of an annual salary in Moscow hotels and restaurant, but are unwilling to give them a fax machine so that they can communicate more easily.

Small talk in Japan

The basis of Japanese business etiquette is mutual trust and respect for individuals, companies and values.

Mind your manners

- The Japanese do not shake hands; making physical contact is an uncomfortable experience. If they offer to shake hands with westerners it should be followed by a slight bow of your head. The handshake should be gentle. Imitate your host.
- Do not look them straight in the eyes. It can appear challenging and makes them feel uncomfortable.
- When you are introduced give them your business card. Cards are also exchanged at parties. You might need up to 30 cards a day. Your cards should be in English and Japanese. You need to have it translated by a qualified professional translator.
- When offered a business card, do not put it in your pocket. Read it immediately and leave it in front of you on the table.
- When talking to the Japanese keep a greater distance than at home. And do not point with your finger, it is impolite.

- Be aware of silences or gaps in the conversation and don't make the western mistake of trying to fill them; silence is »golden« after all.
- They are highly status-conscious. Top level executives expect to do business with top level managers from the other company. It is a sign of respect. Never use first names.
- They enjoy giving carefully chosen personal gifts. Westerners are often surprised at the trouble they take to find out about a visitor's family. Follow their example when they visit you.
- The Japanese Samurai feels uncomfortable in the presence of clever businesswomen who try to negotiate in an aggressive way. Women are rarely invited into the men's social circles.
- Dress conservatively: blue or grey suit, white shirt, dark tie. They like clean, not too young, not too hairy, calm and modest business partners.

Small talk in China

Westerners see China as a Third World underdeveloped country. They forget that it is also the oldest civilisation on earth that invented gun powder and porcelain at a time when Europeans were still throwing stones at each other. They have civilised Japan and all their neighbouring nations. They see themselves as the middle Empire, the centre of the universe, which might become the planet's biggest future market. They do no trust westerners, who in some parts of China are still considered as »foreign devils«.

Be patient when doing business in China – just how patient will be illustrated by the following story.

A young Chinese man was called to his boss who was to retire soon. The old man asked him if he was ready to take his position in the communist party. The young man was overjoyed. »One thing you must remember when dealing with people«, said the old man, »always be patient.« The young man nodded. The old man repeated his advice three times, and when he was going to repeat it a fourth time, the young man said »Do you take me for an idiot? Why do you repeat such a simple thing four times?« And the old man smiled and said »I've said it a few times, and you are already impatient.«

Mind your manners

- Offer gifts in private, not in front of others, as in Arab countries.
- Don't imitate your Chinese partners. Be yourself.
- Don't show off your knowledge of Chinese culture and traditions even if you think you are an expert on Chinese affairs. (By the way, you are not and will never be one.) They might interpret it as western arrogance.

- Be modest. Make excuses that you don't know much about their culture. They will forgive you.
- In China a wise person knows when to be quiet. Don't talk too much.
- When an Asian smiles it could be a signal that he is embarrassed about some bad news.
- If a toast is made, there will be a clapping of hands. Clap back and smile.
- Don't refuse food or drinks. Your host will be offended. Learn to eat with chopsticks. Don't be afraid – it's a good omen if you drop them on the floor.
- Don't refer to China as Mainland China. It is the People's Republic of China. And Taiwan is not a sovereign state but one of the Republic's provinces.
- Learn a few words of Chinese to be able to make polite noises. It's good if you know enough Mandarin to understand what your partners and your interpreter are saying. Don't try to speak it, even if you are under the illusion that you can. They won't appreciate your accent.

In China time is relative, which the following story shows

A Chinese official informed an ARCO manager that tomorrow China would be the number one nation in the world. The American said that he did not doubt it, considering the size of the country and its population, and the tremendous technical progress being made. He asked, »When do you think that China will become number one?« The Chinese official answered: »Oh, in four or five hundred years.«

Small talk in South Korea

Within the last 50 years South Korea has developed from an agricultural country to one of the Tiger Economies. Korean society is highly structured; great respect is paid to age. They are Buddhists with a strong Confucian tradition.

Mind your manners

- When invited to a Korean home remove your shoes and wait to be invited inside.
- Don't knock at the door, it's a western tradition. Koreans find it intrusive. Cough or call out their names instead.
- They are on first-name terms only with people they know very well.
- Avoid topics like socialism, communism or Korean politics.
- At parties stay with your hosts or your group. Don't circulate on your own. At home, don't leave your Korean guests. Circulate with them.
- They tend to stand close to you when talking and touch your arm to attract attention.

- Don't be offended when a Korean asks: »Where are you going? What time are you going to eat? What's that book you're reading?« Smile and say something vague – that's what is expected.
- Be prepared for personal questions: »Are you married? How old are you? How much do you earn?«
- Conversation takes place after the meal, not during the meal.
- After a meal you say »I've eaten very well. It was really delicious.« A mere »Thank you« is not enough. It sounds insincere.
- Smiles are reserved for personal relationships. Smiling at newcomers is considered to be too pushy.
- Laughter is a sign of embarrassment, a sort of apology after having committed a blunder.
- Don't scratch your nose. Finger-to-nose gestures are very rude. Avoid putting your fingers near to your face.
- Saying »no« when offered something is a sign of politeness. Say no a couple of times and then accept. You don't have to eat, drink or smoke it.

Small talk in Arab countries

The 21 states of the Arab League are an interesting market for European companies. These states cover a territory three times the size of Europe and many are rich in oil. Their total population is over 200 million. No Arab constitution is older than 90 years. European managers do not think highly of Arab countries: »Where would they be without their oil?« The Arabs, however, have not forgotten that they were once the leading civilisation in the western world. They try to strike a balance between modern technology and their traditions.

Mind your manners

- First, don't say anything to insult their country. They believe they were world leaders and can be again. Of course we just see them living off oil revenues. Don't say anything like that to them.
- Here in the West we are self-centred. They focus on the family. So if you get the chance, wish members of the family the best. But not any of the females – don't compliment the cooking. Take the women's work for granted. Or you may get into trouble.
- Arabs tend to come very close to you when speaking. They touch a lot as well. Don't move away if possible otherwise they'll feel you're being arrogant.
- Look them in the eye when talking to them. Eye contact counts for a lot. And try praising their country, art, dress and food (but not their women).

- Learn a few polite noises in Arabic.
- Accept all invitations.
- Give your gift in front of others so that there will be the appearance of bribery.
- Don't ask them to unpack a present in your presence.
- Take only the food that is offered to you.
- Eat with your right hand.
- Avoid unpleasant subjects like accidents, poverty, death.
- Leave soon after eating.
- Praise their country, food and fashion.
- Don't refer to the Gulf as the Persian Gulf. It is the Arab Gulf.
- Don't shake hands with women.
- Don't enquire about your host's wives or daughters.
- Don't discuss women, politics or religion.
- Don't admire your host's pictures, furniture or horses. He might feel obliged to give them to you as a present.
- Don't point the sole of the foot towards someone when seated.

You had better read the Koran if you want to do business with them. In the Gulf States a good manager is a good Moslem. He will often refer to Allah and the Koran. Islam influences every aspect in the Arab world especially in Saudi Arabia. The Third Development Plan for the Kingdom of Saudi Arabia, 1980-1985 clearly stated: »The distinguishing mark of the Saudi approach to development is that its material and social objectives are derived from the ethical principles of Islam and the cultural values of Saudi society.« So you and your partners will often have different views of what is right or wrong.

They talk about other things before they talk about business. Then after having talked about business they might come back to social matters and so on. The Arab language is a poetic language, rich in associations and allusions. It is important for native speakers that they say what they mean. They tend to over-express themselves using exaggerations, fantastic metaphors, strings of adjectives and repetition accompanied by vivid body language. To the Arab, loudness is a sign of strength and sincerity, soft tones imply weakness and a lack of sincerity. Remember, the Gulf War took place partly because George Bush spoke softly and Saddam Hussein did not believe what he said about declaring war.

Lexikon der Berufs- und Abteilungsbezeichnungen

In diesem Anhang finden Sie Wörterlisten mit Berufs- und Abteilungsbezeichnungen auf Englisch und Deutsch. Aus Gründen der besseren Lesbarkeit wurde großteils darauf verzichtet, jeweils die weibliche und die männliche Bezeichnung zu verwenden.

Berufsbezeichnungen (Englisch/Deutsch)

Sie können die folgende Liste alphabetisch sortiert nach den deutschen Berufsbezeichnungen auf mybook.haufe.de herunterladen.

English	German
accountant	Buchhalter
accounting manager	Leiter/in Buchhaltung
accounts payable manager	Leiter/in Kreditorenbuchhaltung
accounts receivable manager	Leiter/in Debitorenbuchhaltung
administrative assistant	Verwaltungsassistent
advertising manager	Leiter/in Werbeabteilung
analyst, credit	Kreditanalyst:in
analyst, financial	Bilanzanalyst:in
applications engineer	Anwendungstechniker:in
appraiser, automobile damage	Kfz-Gutachter:in
apprentice	Lehrling
attorney, patent	Patentanwalt:in
auditor	Wirtschaftsprüfer:in
auxiliary worker	Hilfskraft
blue-collar worker	Gewerbliche Arbeitskraft
budget manager	Leiter/in Budgetverwaltung
business unit manager	Leiter/in Geschäftseinheit
CEO (chief executive officer)	Geschäftsführer:in
CFO (chief financial officer)	Leiter/in Finanzwesen
chairman of the executive board	Vorstandsvorsitzender:e
chemical engineer	Verfahrensingenieur:in
chemical technician	Chemietechniker:in
chief engineer	Chefingenieur:in
chief human resources officer	Personalleiter:in
civil engineer	Bauingenieur:in
clerk	Sachbearbeiter:in

English	German
commercial clerk	Bürokaufmann:frau
company doctor	Betriebsarzt/Betriebsärztin
compensation manager	Leiter/in Vergütung und Gehälter
consultant	Berater:in
controller	Kostenprüfer:in
copywriter	Werbetexter:in
country manager	Leiter:in Geschäftsbereich einer ausländischen Niederlassung
department leader	Abteilungsleiter:in
deputy head of …	Stellvertretender:e Leiter:in …
design/development engineer	Entwickler:in
design engineer	Konstruktionsingenieur:in
designer	Konstrukteur:in
director human resources	Leiter:in Personalabteilung
director materials and logistics	Leiter:in Material und Logistik
district manager	Bezirksleiter:in
driver	Fahrer:in
electrician	Elektriker:in
employee development manager	Leiter:in Personalentwicklung
executive	Führungskraft
executive board member	Vorstandsmitglied
executive secretary	Chefsekretär:in
financial analyst	Bilanzanalyst:in
fitter	Monteur:in
foreman	Vorarbeiter:in
graphic designer	Grafiker:in
head of …	Leiter:in …
head of advertising	Leiter:in der Werbeabteilung
head of personnel	Personalleiter:in
industrial engineer	Wirtschaftsingenieur:in

English	German
intern	Praktikant:in
investment analyst	Wertpapieranalyst:in
key account manager	Großkundenbetreuer:in
lab assistant	Laborassistent:in
lab technician	Labortechniker:in
line manager	Vorgesetzter:e
maintenance technician	Wartungsmechaniker:in
management trainee	Trainee
manager product development	Leiter:in Produktentwicklung
managing clerk	Disponent:in
manufacturing manager	Produktionsleiter:in
office assistant	Büroassistent:in
office manager	Büroleiter:in
operations manager	Betriebsleiter:in
part-time employee	Teilzeitbeschäftigter:e
personnel clerk	Personalsachbearbeiter:in
personnel development consultant	Berater:in für Personalentwicklung
plant manager	Werkleiter:in
press spokesman	Pressesprecher:in
product manager	Produktverantwortlicher:e
production engineer	Fertigungsingenieur:in
programmer	Programmierer:in
project leader	Projektleiter:in
proofreader	Korrekturleser:in
quality assurance manager	Leiter:in Qualitätssicherung
quality control manager	Leiter:in Qualitätskontrolle
receptionist	Empfangsmitarbeiter:in
sales clerk	Verkaufssachbearbeiter:in
sales manager	Vertriebsleiter:in
sales representative	Außendienstmitarbeiter:in

English	German
security guard	Pförtner:in
semi-skilled worker	angelernte Arbeitskraft
SHE manager (safety, health and environment protection manager)	Leiter:in Arbeitssicherheit und Umweltschutz
skilled worker	Facharbeiter:in
store foreman	Lagermeister:in
strategic planning manager	Leiter:in strategische Planung
supervisory board	Aufsichtsrat:in
technical manager	Technischer:e Leiter:in
temporary worker	Zeitarbeiter:in
trainee	Auszubildender:e
trainer	Ausbilder:in
training manager	Ausbildungsleiter:in
translator	Übersetzer:in
unskilled worker	Hilfsarbeiter:in
white-collar worker	Angestellter:e
works council	Betriebsrat:in

Abteilungsbezeichnungen (English/Deutsch)

Sie können die folgende Liste, alphabetisch sortiert nach den deutschen Abteilungsbezeichnungen, auf mybook.haufe.de herunterladen.

English	German
Accounting	Buchhaltung
Administration Department	Verwaltung
Assembly	Montage
Complaints Department	Beschwerdeabteilung
Controlling	Controlling
Customer Service	Kundenservice
Design Department	Konstruktion

English	German
Dispatch/Shipping Department	Versandabteilung
Finance Department	Finanzabteilung
Human Resources	Personalabteilung
Incoming Goods Department	Wareneingang
IT/Information Technology Department	Informationstechnik
Legal Department	Rechtsabteilung
Management	Geschäftsleitung
Market Research	Marktforschung
Marketing Department	Marketingabteilung
Office Supplies	Materialbeschaffung
Patent Department	Patentabteilung
Payroll Accounting	Lohnbuchhaltung
Personnel Development	Personalentwicklung
Post Room	Poststelle
PR/Public Relations	PR/Öffentlichkeitsarbeit
Press Office	Presseabteilung
Production	Herstellung
Purchasing Department	Einkaufsabteilung
Quality Assurance	Qualitätssicherung
Recruitment	Personalbeschaffung
Research & Development	Forschung u. Entwicklung
Sales Department	Verkaufsabteilung
Sales Representatives	Außendienst
Training	Ausbildung/Weiterbildung
Warehouse	Lagerhalle

Glossar gebräuchlicher Abkürzungen

Auch im Bereich Human Resources begegnet man häufig Abkürzungen, deren Bedeutung nicht immer gleich offensichtlich ist. Auf den folgenden Seiten finden Sie daher eine Zusammenstellung solcher Abkürzungen und deren Bedeutung.

Abbreviation	Meaning
AAP	Affirmative Action Plan
AAR	After Action Review
ABM	Activities Based Management
AE	Account Executive
ANSI	American National Standards Institute
AP	Accounts Payable
AR	Accounts Receivable / Annual Review
AVC	Additional Voluntary Contributions
AWOL	Absent Without Leave
B2B	Business to Business
B2C	Business to Consumer
BA	Bachelor of Arts
BCP	Business Continuity Plan
BOD	Board of Directors
B. Sc.	Bachelor of Science
BST	Basic Skills Trainer
C&B	Compensation and Benefits
CAI	Computer Assisted Instruction
CBL	Computer Based Learning
CBT	Computer Based Training
CEO	Chief Executive Officer
CFO	Chief Financial Officer
CHRO	Chief Human Resources Officer
CIO	Chief Investment Officer, Chief Information Officer
COB	Close of Business
COLA	Cost of Living Allowance
COO	Chief Operating Officer
CPD	Continuing Personal Development
CPE	Continuing Professional Education
CRE	Commission for Racial Equality
CRO	Companies Registration Office

Abbreviation	Meaning
CSO	Chief Security Officer
CTO	Compensatory Time Off
CV	Curriculum Vitae
D&O	Directors and Officers
DCPP	Defined Contribution Pension Plan
DINKS	Double Income No Kids
DOB	Date of Birth
DOI	Date of Injury
DRP	Disaster Recovery Plan
EAP	Employee Assistance Programme
EBO	Employee Buyout
EBT	Earnings Before Taxes
EDP	Electronic Data Processing
EE	Employee
EEO	Equal Employment Opportunity
EHO	Environmental Health Officer
EIN	Employer Identification Number
EOC	Equal Opportunities Commission
EOD	End of Day
EOY	End of Year
EPA	Equal Pay Act
ESOP	Employee Stock Option Plan
EWC	European Works Council
FAQ	Frequently Asked Questions
FEP	Fair Employment Practice
FOJT	Formal on the Job Training
FT	Full Time
GMP	Guaranteed Minimum Pension
GPHR	Global Professional in Human Resources
HF	Human Factors
HPT	Human Performance Technology

Abbreviation	Meaning
HR	Human Resources
HRD	Human Resources Development
HRIS	Human Resources Information System
HRM	Human Resources Management
HRMS	Human Resources Management System
HSC	Health & Safety Commission
HUMINT	Human Intelligence
I&D	Integration and Development
ILO	International Labour Organisation
ILO	International Labour Organisation
IMI	Interactive Multimedia Instruction
INSET	In-service Training
IR	Industrial Relations
ITP	Individual Training Plan
JD	Job Description / Juris Doctorate
KISS	Keep It Simple, Stupid
KPI	Key Performance Indicator
KPM	Key Performance Measures
KSA	Knowledge, Skills & Abilities
LMS	Learning Management System
LOA	Leave of Absence
LOS	Length of Service
LR	Labor Relations
LTD	Long-Term Disability
LWOP	Leave Without Pay
LWP	Leave with Pay
M&A	Merger & Acquisition
MBA	Master of Business Administration
MBO	Management by Objectives / Management Buyout
MOP	Measure of Performance
MPP	Maternity pay period

Abbreviation	Meaning
MQ	Minimum Qualifications
NE	Non-Exempt
NI	National Insurance
NMW	National Minimum Wage
NVQ	National Vocational Qualification
O&M	Organisation and Methods
OAP	Old Age Pensioner
OD	Organization Development
OE	Operating Expense
OI	Operating Income
OJT	On-the-Job-Training
OSHA	Occupational Safety and Health Administration
OT	Overtime
p. a.	per annum
PA	Performance Assessment / Personal Assistant
PAYE	Pay As You Earn
PBR	Payment by Results
PHR	Professional in Human Resources
PIP	Performance Improvement Plan
PIW	Period of Incapacity for Work
PM	Performance Management / Project Management
PPP	Personal Pension Plan
PRP	Performance-related Pay / Profit-related Py
PT	Part Time
QR	Quarterly Review
QWL	Quality of Working Life
R&R	Ranking and Rating
RTW	Return to Work
SAYE	Save As You Earn
SBA	Small Business Administration
SERPS	State Earnings-related Pension Scheme

Abbreviation	Meaning
SHRM	Society for Human Resources Management
SMP	Statutory Maternity Pay
SPHR	Senior Professional in Human Resources
SSP	Statutory Sick Pay
STD	Short Term Disability
SWOT	Strengths, Weaknesses, Opportunities, and Threats
T&D	Training and Development
TBD	To Be Done / To Be Discussed
TBT	Technology-based Training
TDB	Temporary Disability Benefits
TM	Talent Management
TNA	Training Needs Analysis
TOIL	Time off in Lieu
TPD	Temporary Partial Disability
TQM	Total Quality Management
TRADE	Training Devices
TTD	Temporary Total Disability
UCI	Unemployment Compensation Insurance
VOC-REHAB	Vocational Rehabilitation
VP	Vice President
WBT	Web-based Training
WIP	Work in Progress
WPS	Work Performance Standards
WTO	World Trade Organization
XRA	Expected Retirement Age
YOYO	You're on Your Own
YTD	Year to Date
ZBB	Zero-based Budgeting

Praktische Phrasen und Redewendungen im Personalwesen

Nachfolgend finden Sie zahlreiche nützliche Phrasen und Redewendungen für Ihre Arbeit als Personaler.

General phrases and idioms in human resources

English	German
be a dab hand at something	ein goldenes Händchen für etwas haben
be fair to middling	so lala, mittelprächtig sein
be head and shoulders above the rest	den anderen haushoch überlegen sein
fall down on the job	nicht die Erwartungen erfüllen
get one's act together	sich zusammenreißen
get one's hand in	sich eingewöhnen
get the axe	gefeuert werden
get the hang of something	den richtigen Dreh finden
give a golden handshake	eine Abfindung zahlen
give someone the boot/sack/chop	jemanden feuern
heads will roll!	da werden Köpfe rollen!
hiring and firing	Einstellen und Entlassen
her heart is not in her work	sie ist mit dem Herzen nicht bei der Sache
If you pay peanuts you get monkeys	von nichts kommt nichts
land a plum job	einen Traumjob ergattern
learn the ropes	die Kniffe lernen
make the grade	das Niveau erreichen
mention the perks	die Leistungsanreize erwähnen
not be up to the mark	nicht das Zeug für die Stelle haben
short-list applicants	Bewerber in die engere Auswahl nehmen
throw pearls to the swine	Perlen vor die Säue werfen
you can't make a silk purse out of a sow's ear	Aus einem Kieselstein kann man keinen Diamanten schleifen

Personalabteilung

English	German
Courses are not to take place in core time.	Kurse sollen nicht in der Kernzeit stattfinden.
define training requirements	den Bedarf an Fortbildung ermitteln
do advanced training	Schulungen für Fortgeschrittene durchführen
hold courses in flexitime	Kurse während der Gleitzeit abhalten
identify strong/weak points	Stärken/Schwächen identifizieren
offer hands-on training	praktische Ausbildung anbieten
pass with distinction	hervorragend abschneiden
provide/give training	Schulungen anbieten/durchführen
take/sit/pass/fail an exam	eine Prüfung ablegen/nicht bestehen
What's the drop-out rate?	Wie hoch ist die Zahl der Abbrecher?

In-house training

English	German
allocate/perform tasks	Aufgaben zuweisen/erledigen
attend a seminar	an einem Seminar teilnehmen
do a job rotation	Rotation am Arbeitsplatz machen
do a placement	ein Praktikum machen
do an apprenticeship	eine (gewerbliche) Lehre machen
do on-the-job-training	am Arbeitsplatz ausgebildet werden
get a diploma	ein Diplom erhalten
get a good report	ein gutes Zeugnis erhalten
land a job as a management trainee	eine Stelle als Trainee bekommen
learn by doing	in der Praxis lernen

Learning after hours

English	German
apply for a scholarship	sich für ein Stipendium bewerben
believe in life-long learning	an das lebenslange Lernen glauben

English	German
cram for an exam	für eine Prüfung büffeln
do a distance-learning course	einen Fernlehrgang absolvieren
enrol at university	sich an einer Universität einschreiben
enrol for a part-time course	sich für einen Teilzeitkurs anmelden
get a degree	mit einem Diplom abschließen
get a scholarship	ein Stipendium bekommen
get exam jitters	Prüfungsangst bekommen
I'm training to be a …	ich mache eine Ausbildung zum …
join an adult education course	einen Volkshochschulkurs besuchen
qualify for a job	sich für einen Posten qualifizieren
register for a test	sich für eine Prüfung anmelden

Evaluating candidates

English	German
analyse someone's performance	jmds Leistung bewerten
appraise/assess/evaluate employees	Mitarbeiter beurteilen/bewerten
expect something of somebody	etwas von jemandem erwarten
he's hard-working	er ist fleißig
hold an employee appraisal	eine Mitarbeiterbeurteilung durchführen
make a rating/performance appraisal	eine Leistungsbewertung durchführen
perform up to par	auf der Höhe sein
sift out the best candidates	die besten Kandidaten aussieben
test trainees	Auszubildende prüfen

The role of the personnel manager

English	German
coach members of staff	Mitarbeiter betreuen
convey/impart knowledge	Wissen vermitteln
demand interpersonal skills	zwischenmenschliche Fertigkeiten fördern
get feedback	Rückmeldungen bekommen

English	German
give a pep talk	jemanden aufmuntern
improve managerial performance	Führungsfähigkeiten verbessern
praise employees	Mitarbeiter loben
select potentials	Potenzialträger auswählen

Building a career

English	German
be a high flyer	ein Überflieger sein
climb the career ladder	in der Hierarchie aufsteigen
cut someone's career short	jemandes Karriere beenden
do a lot of overtime	viele Überstunden machen
do the job of one's life	ein Meisterstück liefern
fall down on the job	bei einer Aufgabe versagen
get promotion	befördert werden
have a good track record in HR	gute Leistungen im Bereich Personalwesen haben
have an outstanding/impressive CV	einen erstklassigen/beeindruckenden Lebenslauf haben
he has get up and go	Unternehmensgeist haben
headhunt somebody	jemanden abwerben
launch/start a career	eine Karriere starten
they are ten a penny	es gibt sie wie Sand am Meer

Salaries and wages

English	German
be appointed CEO	zum CEO ernannt werden
be in a salary group/grade	einer Gehaltsgruppe angehören
be on the payroll of the company	auf der Gehaltsliste einer Firma stehen
do piece work	im Akkord arbeiten
get a negotiated salary	außertarifliches Gehalt beziehen
get an advance	einen Lohnvorschuss erhalten

English	German
hand in one's tax return	eine Einkommenssteuererklärung einreichen
have employees on the payroll	Mitarbeiter auf der Gehaltsliste haben
introduce a wage freeze	die Löhne einfrieren
offer adequate remuneration	angemessene Vergütung anbieten
pay income tax	Lohnsteuer bezahlen
receive incentives and benefits	Leistungszulagen und Vergünstigungen erhalten
work as a wages clerk	als Lohnsachbearbeiter arbeiten

Getting paid

English	German
be paid overtime	Überstunden bezahlt bekommen
belong to an old age pension scheme	eine Altersversorgung haben
contribute to a company pension plan	Beiträge in eine betriebliche Altersversorgung leisten
draw a company pension	Firmenrente beziehen
earn a bonus	eine Prämie verdienen
get a company flat/company car	eine Firmenwohnung/Auto bekommen
get paid time and a half	jmd 50 % Zuschlag bezahlen
have a 9 to 5 job	ein Arbeitsplatz mit geregelter Arbeitszeit haben
join the company's medical insurance	in die Betriebskrankenversicherung eintreten
profit-sharing for executives	Gewinnbeteiligung für leitende Angestellte
right to paid holiday	Anrecht auf bezahlten Urlaub
the salary package includes …	das Gehaltspaket beinhaltet …

Negotiating a salary

English	German
agree an hourly wage	einen Stundenlohn vereinbaren
ask for a raise	um Gehaltserhöhung bitten
bear the costs of relocation	den Umzug bezahlen
demand top wages	Spitzenlöhne verlangen

English	German
get the minimum pay rate/wage	den Mindestlohn erhalten
negotiate the salary	das Gehalt aushandeln
pay the collective agreement	Tariflohn/-Gehalt zahlen

The legal side

English	German
be on a trial/probationary period	auf Probe eingestellt werden
break a contract	einen Vertrag brechen
draft/draw up a contract/agreement	einen Anstellungsvertrag aufsetzen
fix the salary grade	die Gehaltsgruppe festsetzen
fulfil an agreement	eine Vereinbarung erfüllen
get a limited/temporary contract	einen befristeten Arbeitsvertrag erhalten
give a period of three months' notice	eine Kündigungsfrist von drei Monaten geben
make an amendment/change	einen Vertrag abändern
sign an agreement/contract	eine Vereinbarung / einen Vertrag unterschreiben
stipulate in the contract	im Vertrag festlegen

Complaints

English	German
be sick to the back teeth about it	die Schnauze voll haben
come out on a wild-cat strike	in einen wilden Streik treten
down tools	die Arbeit niederlegen
expect a reward for one's work	eine Belohnung für seine Arbeit erwarten
feel underpaid	sich unterbezahlt fühlen
He doesn't have a penny to his name.	Er besitzt keinen Pfennig.
live beyond one's means	über seine Verhältnisse leben
take time off sick	sich krank melden
work-to-rule	Dienst nach Vorschrift machen
You have to bend over backwards.	Du musst dir allergrößte Mühe geben.

The employer's point of view

English	German
be a hard nut to crack	eine harte Nuss zu knacken geben
be a shooting star	ein Senkrechtstarter sein
be on the short-list	in der engeren Auswahl stehen
bring in a wage freeze	einen Lohnstopp einführen
cancel the Christmas bonus	Weihnachtsgeld streichen
contact a headhunter	einen Personalvermittler kontaktieren
fill a vacancy	eine freie Stelle besetzen
good workers are far and few between	gute Arbeitskräfte sind spärlich gesät
guarantee a minimum wage	einen Mindestlohn garantieren
I want to moonlight/go moonlighting	ich will schwarzarbeiten
offer basic rates of pay	Grundlohn bezahlen
pay an attractive wage/salary	ein reizvolles Gehalt bezahlen
pull one's socks up	sich am Riemen reißen
reduce absenteeism	Fehlzeiten abbauen
The love of money is the root of all evil.	Die Liebe zum Geld ist die Wurzel allen Übels.
this team is not up scratch	Dieses Teamist nicht auf Zack
We expect hard work and loyalty.	Wir verlangen harte Arbeit und Loyalität.
We're not a horn of plenty.	Wir sind kein Füllhorn.
You never had it so good!	Dir ging es nie so gut wie jetzt!

Industrial relations

English	German
a line of pickets	eine Streikpostenlinie
be a member of a trade union	Gewerkschaftsmitglied sein
be a member of the Works Council	ein Mitglied des Betriebsrats sein
be on a go-slow	im Bummelstreik sein
be out of work/unemployed/jobless	arbeitslos sein
call a strike	zum Streik aufrufen
cross the picket line	Streikpostenlinie durchbrechen

English	German
demand better working conditions	bessere Arbeitsplatzbedingungen fördern
down tools	Arbeit niederlegen
give notice/resign	kündigen
go on strike	streiken
look for a job	einen Arbeitsplatz suchen
reduce the unemployment rate	die Arbeitslosenrate reduzieren
work-to-rule	Arbeit nach Vorschrift

Becoming self-employed

English	German
be a free-lancer	Freiberufler sein
be in competition with somebody	mit jemandem konkurrieren
be self-employed	sich selbstständig machen
He must paddle his own canoe.	auf eigenen Füßen stehen
jump in at the deep end	ins kalte Wasser springen
keep one's end up	seinen Teil beitragen
rise from the ranks	sich hocharbeiten
run a business	einen Betrieb leiten
take a big risk	ein großes Risiko eingehen
take over/acquire a company	eine Firma übernehmen
work on a free-lance basis	freiberuflich tätig sein
work on your own account	auf eigene Rechnung arbeiten
You must pull your weight.	Du musst deinen Teil beitragen.

Wörterbuch Englisch-Deutsch

Nachfolgend finden Sie eine umfassende Zusammenstellung der wichtigsten fachspezifischen Vokabeln, die in diesem Buch verwendet werden.

English	German
A	
A level (= Advanced Level)	Abitur
abbreviation	Abkürzung
ability	Fähigkeit
above average	überdurchschnittlich
absence	Abwesenheit
absenteeism	Abwesenheit; Fehlzeiten
abuse	Missbrauch
academic record	akademische Bildung
access to	Zugriff auf
access, get	Zutritt erhalten
accident prevention	Unfallverhütung
accommodate	unterbringen
accompany	begleiten
accomplished	geschafft, fertig gestellt, vollendet
accounts ledger	Kontenplan
accrue	angefallen, aufgelaufen
accuracy	Genauigkeit
accurate	genau
achieved	erreicht
achievement	Leistung
acquire	erwerben, erlangen
adaptable	anpassungsfähig
add oil to the flames	Öl ins Feuer gießen
adequate	angemessen
admirer	Bewunderer
admit guilt	Schuld eingestehen
admonish	abmahnen; ermahnen
advantage	Vorteil
advertisement	Stellenanzeige

English	German
advertising	inserieren, Werbung
advised	beraten
age	Alter
age limit	Altersbeschränkung
ageism	Altersdiskriminierung
agency	Agentur
agreement	Vereinbarung
amend	ändern; überarbeiten
analysed	analysiert
analytical abilities	analytische Fähigkeiten
announce	bekanntgeben
annual	jährlich
anticipate	erwarten
appeal; to	anfechten
appeal	Einspruch
appearance	äußeres Erscheinungsbild
applicant	Bewerber
application	Bewerbung
appoint	anstellen; ernennen
appointed	ernannt, berufen
appointment; keep an ~	einen Termin einhalten
appraisal	Bewertung, Beurteilung
appraisal interview	Mitarbeitergespräch
appraised	bewertet
appreciate	dankbar sein
appreciate something	etwas schätzen
apprenticeship	Lehre
approach	Ansatz
appropriate	passend, geeignet
approved	freigegeben, genehmigt, zugelassen

English	German
arbitrate	schlichten
arbitration	Schlichtung
assess	bewerten
assessor	Gutachter
asset for your company	Gewinn für Ihr Unternehmen
assumption	Annahme
assurance	Gewissheit; Sicherheit
attachment of salary	Lohnpfändung
attendance	Anwesenheit, zugegen sein
attention	Aufmerksamkeit
attitude	Einstellung
audience	Publikum, Zuhörer
authoritative	bestimmend
available	verfügbar
average	durchschnittlich
avoid	vermeiden
award	auszeichnen
award a degree	einen akademischen Grad verleihen
awareness	Bewusstsein
B	
Baccalaureate	Hochschulreife
bank clerk	Bankkaufmann /-frau
bank transfer	Banküberweisung
basic award	Grundentschädigung
basically	grundsätzlich/im Grunde
battery of activities	eine Reihe von Aktivitäten
behalf; on ~ of	im Namen von
behave	verhalten, sich benehmen
behaviour	Verhalten
belief	glauben, vertrauen

English	German
benefits	Arbeitgeberleistungen
bereavement	Trauerfall
best of luck	viel Glück
beware of	sich in Acht nehmen; Vorsicht!
biased	befangen; parteiisch; voreingenommen
bi-lingual	zweisprachig
blue-collar worker	Arbeiter
body language	Körpersprache
bottom line; the ~ is	im Endeffekt, unterm Strich
bought	gekauft
boycott	Boykott
break off talks	Gespräche abbrechen
break-down	Aufschlüsselung
brief (n)	Kurzdarstellung
briefly	kurz
bring somebody down to earth	jemand auf den Boden der Tatsachen zurückholen
built	gebaut, aufgebaut
bullet point	Aufzählungspunkt
bully	mobben
bully somebody	jmdn. schikanieren, einschüchtern, mobben
bullying	Schikane, Mobbing
business administration	Betriebswirtschaft
but beware	aber Achtung
C	
calm down	sich beruhigen
candidate	Bewerber:in, Kandidat:in
capable	fähig
capacity; in his ~	in seiner Funktion
capacity; in the ~ of	in der Eigenschaft als
career change	Berufs-/Karrierewechsel

English	German
career path	berufliche Laufbahn
carefully	sorgfältig
case study	Fallstudie
cashier	Kassierer
cause waves	Wellen schlagen
celebrate	feiern
certificate	Abschluss; Zertifikat
cession	Abtretung
challenge	Herausforderung
challenging	herausfordernd
character trait	Charaktereigenschaft
charge with	beauftragen
check something	etwas überprüfen
cheerful	fröhlich, heiter
chronicles	Chronik, Aufzeichnungen
circumstance	Umstand
civil servant	Beamter
clarify	klären
clean driving licence	Führerschein ohne Verkehrsregistereinträge
clerical staff	Büropersonal
closed shop	Betriebe, in denen die Beschäftigten einer Gewerkschaft angehören müssen
clue	Anhaltspunkt
code of conduct	Verhaltenskodex
coercion	Zwang
cognitive	kognitiv
collective bargaining	Tarifverhandlungen
collective bargaining agreement	Tarifvereinbarung
combat crime	Kriminalität bekämpfen
combined	kombiniert

English	German
come straight to the point	gleich auf den Punkt kommen
comment	Kommentar
commercial clerk	Sachbearbeiter:in
commit to something	sich zu etwas verpflichten
commitment	Einsatz; Engagement
commitment; personal ~	persönlicher Einsatz
common; to have something in ~	etwas gemeinsam haben
communication skills	kommunikative Fähigkeiten
company car	Firmenwagen
company objectives	Unternehmensziele
company pension	Betriebsrente
compassion	Mitgefühl
compensation	Abfindung
compensation; financial ~	Abfindung
competence	Fähigkeit; Kompetenz
competitive	konkurrenzfähig
competitor	Konkurrent
compile	zusammenstellen
complaint	Beschwerde
completed	erledigt, vollendet, fertig gestellt
compliment	Lob, Anerkennung
comply with	befolgen
compromise	Kompromiss
computer literate	über gute Computerkenntnisse verfügen
concern	Bedenken
concerning	bezüglich, betreffend
conclusion	Schlussfolgerung
condition	Bedingung
conditions of employment	Arbeitsbedingungen
condolence	Beileid

English	German
conduct	Handlungsweise, Verhalten
conducted	durchgeführt, abgewickelt
confidential; keep ~	vertraulich behandeln
confidentiality	Vertraulichkeit
confirm	bestätigen
conflict	Konflikt
conflict of interest	Interessenkonflikt
consent	Zustimmung
consideration; give ~ to something	etwas in Erwägung ziehen
consist of	umfassen; beinhalten
constructive dismissal	konstruktive Kündigung
constructive suggestion	konstruktiver Vorschlag
construe	auslegen
contact details	Kontaktdaten
contagious	ansteckend
continuous service	ununterbrochene Firmenzugehörigkeit
contract condition	Vertragsbedingung
contract of employment	Arbeitsvertrag
contract; conclude a ~	einen Vertrag abschließen
contribution	Beitrag
contributor	Mitarbeiter
controlled	kontrolliert
convictions	Vorstrafen
convincing	überzeugend
cope with	zurechtkommen mit, etwas bewältigen, meistern
copywriting	Texten
core business	Kerngeschäft
corporate clients	Firmenkunden
counsel	beraten
counter-productive	kontraproduktive

English	German
course of study	Studium
covering letter	Anschreiben
coward	Feigling
co-worker	Kollege
created	erstellt, geschaffen
criterion, criteria	Kriterium, Kriterien
cross one's mind	jmdn. in den Sinn kommen
cross somebody's mind	jemandem in den Sinn kommen; einfallen
cross the picket line	Streikposten durchbrechen
curriculum vitae (CV)	Lebenslauf
curve	Kurve
cushion	abfedern
customer's confidence	Kundenvertrauen
cut	eingespart, gekürzt
D	
data protection	Datenschutz
day-to-day	täglich
deadline	Abgabefrist
deadline; ensure a ~	Einhaltung eines Termins sicherstellen
deadline; meet the ~	Frist einhalten
debriefing	Abschlussbesprechung
decision	Entscheidung
decision-making	Entscheidungsfindung
decision-making process	Entscheidungsprozess
deduction	Abzug
deem	etwas befinden; erachten als
deemed; shall be ~ as	gelten als
deepest sympathy	tiefstes Mitgefühl
defaulter	Schuldige(r)
degree	Hochschulabschluss, Abschluss

English	German
delegate	delegieren
delivered	gelieferte
demand	Nachfrage
demonstrate	unter Beweis stellen
demonstrated	aufgezeigt, demonstriert
denial	Ableugnen, Bestreiten
deny	bestreiten, leugnen
department	Abteilung
deserve a special mention	eine besondere Erwähnung verdienen
designed	entworfen
desperate	verzweifelt
detailed testimonial	qualifiziertes Zeugnis
determine	bestimmen
detrimentally	schädlich, nachteilig
developed	entwickelt
development; personnel ~	Personalentwicklung
diploma	Diplom
directed	geführt, geleitet
disability	Behinderung
discharge	Entlassung
disclosure	Offenlegung
discourage (v)	entmutigen
discretion; at the ~ of	im Ermessen von
discriminate	diskriminieren
discriminatory	diskriminierend
dismiss	verwerfen; ablehnen
dismissal	Kündigung, Entlassung
disregard	ignorieren, missachten
dissatisfaction	Unzufriedenheit
distance learning	Fernstudium

English	German
distinction	Auszeichnung
distribute	verteilen
distributors	Händler
disturb	stören
disturbance	Störung
DOB (date of birth)	Geburtstag
document; to ~ something	etw. dokumentieren
documentation	Dokumentation
donate	spenden
draw a line under something	einen Schlussstrich unter etwas ziehen
draw attention to oneself	sich in Szene setzen
draw attention to something	Aufmerksamkeit auf etwas lenken
drawn-out	langgezogen
drove	gelenkt, angetrieben
due	fällig
due and timely manner	ordnungsgemäß und fristgerecht
duty	Aufgabe, Pflicht
E	
economic downturn	Konjunkturabschwung, wirtschaftlicher Rückgang
economics	Volkswirtschaft
education	Bildung
educational leave	Bildungsurlaub
effect; with ~ from	mit Wirkung vom
effort	Aufwand
effort; make an ~	sich bemühen
ejector seat	Schleudersitz
embarrassment	Verlegenheit
empathise with somebody	sich in jmdn. einfühlen;
empathy	Empathie; Einfühlungsvermögen
emphasise	betonen

English	German
employed	eingesetzt, eingestellt
employee security	Mitarbeitersicherheit
employees' rights	Arbeitnehmerrechte
employer-employee relationship	Arbeitsverhältnis
employment contract	Arbeitsvertrag
encounter	begegnen; antreffen
encourage	ermutigen
endorsement	Indossament
engineered	entwickelt
enhance	verbessern
enlarge	vergrößern
enquiry	Anfrage
entertainment allowance	Bewirtungsspesen
entitled	Anspruch haben
entitlement	Anspruch
entrepreneurial and strategic thinking	unternehmerisches und strategisches Denken
environment	Umfeld; Umwelt
equal opportunities	Chancengleichheit
equal opportunities policy	Richtlinie zur Chancengleichheit
equal opportunity	Chancengleichheit
establish	festlegen
esteemed	angesehen; geschätzt
evaluate	bewerten
evaluation standard	Bewertungsmaßstab
evidence	Beweis
evil	Übel
examined	untersucht
exceed	übertreffen
exceeded our high expectations	unsere hohen Erwartungen übertroffen
excess; in ~ of	über; höher als

English	German
exclude	ausschließen
expatriate staff	Mitarbeiter, die ins Ausland entsendet wurden
expect the worst	vom Schlimmsten ausgehen
expectation	Erwartung
expire	auslaufen, enden
express an opinion	seine Meinung kundtun
externally	extern
eye contact	Augenkontakt
F	
face something	mit etwas rechnen müssen
factual	sachlich
false beginner	Anfänger mit geringen Vorkenntnissen
favour (v)	bevorzugen
favour; in ~ of	für etwas sein
fee	Gebühr
feel inconvenienced	sich belästigt fühlen
feud	Streit; Fehde
file a complaint	eine Beschwerde einreichen
filing system	Ablagesystem
fill the gap	die Lücke schließen
findings	Erkenntnisse
fine-tune	fein abstimmen
finger-tapping	Fingertippen
fit the job	dem Anforderungsprofil entsprechen
flowchart	Ablaufdiagramm
fluent	fließend
folder	Ordner; Mappe
follow the rules	Regeln befolgen
follow-up	im Anschluss an etwas
foolish	dumm

English	German
force	zwingen
forecast (v)	vorhergesagt, prognostiziert
forecast	Vorhersage
formed	gestaltet
foster	unterstützen, fördern
framework; within the ~ of	im Rahmen von
further education	Weiterbildung
G	
gain	Bereicherung
gap year	Zeit ohne Beschäftigung/Auszeit
gather	sammeln
gathering	Zusammentreffen; Zusammenkunft
GCE, A level (= General Certificate of Education, Advanced Level)	Abitur
GCSE (= General Certificate of Secondary Education)	Mittlere Reife
gender	Geschlecht
generated	erstellt, generiert
get away with …	davonkommen
get control of oneself	sich beherrschen
get down to business	zur Sache kommen; eine Aufgabe anpacken
get promotion	befördert werden
get to the bottom of	Dingen auf den Grund gehen
glory	Herrlichkeit; Ruhm
go down well; it doesn't ~	das kommt nicht gut an
goal	Ziel
goodness sake; for ~	um Himmels Willen
governing law	geltendes Recht
graduate	Absolvent, Hochschulabsolvent
grasp	Verständnis
grasp something	etwas verstehen, begreifen

English	German
grateful	dankbar
greet	grüßen
grievance	Beschwerde
grievance procedure	Beschwerdeverfahren
gross	brutto
gross salary	Bruttogehalt
guidance	Führung; Anleitung
guided	geführt, geleitet
H	
handle; get a ~ on something	etwas in den Griff bekommen
handout	Handout, Tischvorlage, Informationsblatt
hands-on knowledge	Praxiswissen
harass	belästigen
harassment	Belästigung
hardly	kaum
head of department	Abteilungsleiter:in
head something	etwas leiten
headed	angeführt, geleitet
head-hunting	Abwerbung
head-on	sich etwas stellen
headquarters	Hauptsitz
health and safety policy	Gesundheits- und Sicherheitsrichtlinie
hearing	Anhörung
heated	in Wallung geraten
hesitate	zögern
hierarchical	hierarchisch
high-flying speech	hochtrabende Rede
hint at something	etwas andeuten
hiring	Einstellen
hold talks	Gespräche führen

English	German
holiday	Urlaub
holiday entitlement	Urlaubsanspruch
holiday pay	Urlaubsgeld
honorary guest	Ehrengast
honoured	geehrt
hostility	Feindschaft
I	
ice-breaker	Eisbrecher
identified	identifiziert
illegal	gesetzeswidrig
illness	Krankheit
image of oneself / self-image	Selbstverständnis
immediately	sofort
impart	mitteilen
impartial	unparteiisch; objektiv
impartiality	Unbefangenheit
implement	umsetzen, anwenden
imply	bedeuten; voraussetzen
impose	auferlegen
impression	Eindruck
Improved	verbessert
improvement	Verbesserung
improvise	improvisieren
improvised	improvisiert
in charge	das Sagen haben
in principle	prinzipiell
inaccuracy	Ungenauigkeit
inaccurate	ungenau
inappropriate	unangemessen
incentive	Anreiz

English	German
include	einschließen
increased	gesteigert, erhöht
incumbent	Stelleninhaber
indefinite period	unbefristet
indicate	auf etwas hindeuten
inductee	neuer:e, einzuweisender:e Mitarbeiter:in
induction	Einführung
initiated	initiiert
innovative	innovativ
insight	Einblick
insist on	bestehen auf
inspired	inspiriert
instigate	anregen
instigated	angeregt
instruction	Anweisung
insubordination	Befehlsverweigerung
insurance clerk	Versicherungssachbearbeiter
insurance; health ~	Krankenversicherung
insurance; private medical ~	private Krankenversicherung
integral	integral; wesentlich
integration abilities	Fähigkeiten sich zu integrieren
intellectual property	geistiges Eigentum
intermediate knowledge	Mittelstufenkenntnisse
internal	intern
intervene	eingreifen
intervention	Intervention; Eingreifen
interview (n)	Bewerbungsgespräch
in-tray exercise	Postkorbübung
invalid	ungültig
invaluable	äußerst wertvoll

English	German
invention	Erfindung
investigated	untersucht
isolated	isoliert; abgegrenzt
issue (v)	ausgeben
issue	Angelegenheit; Problem
J	
job advert/advertisement	Stellenanzeige
job description	Stellenbeschreibung, Arbeitsplatzbeschreibung
job offer	Stellenangebot
job performance	Arbeitsleistung
job title	Stellenbezeichnung
job-hunting	auf Arbeitssuche sein
job-related questions	berufsbezogene Fragen
job-seeker	Arbeitssuchender
K	
keep a record	Protokoll führen
keep something in mind	etwas bedenken; berücksichtigen
L	
labels; attach ~ to	in eine Schublade stecken
last	dauern; anhalten
laudable	lobenswert
laugh about something	über etwas lachen
launched	eingeführt
layout	Ausgestaltung
leadership	Mitarbeiterführung
leadership performance	Führungsleistung
leading edge	führend in etwas sein
leave	Urlaub
led	geführt, geleitet
liaise with	sich abstimmen, Verbindung aufnehmen mit

English	German
liaison	Verbindung, Kontakt, Zusammenarbeit
likewise	ebenso
link	Verbindung
loan	Darlehen
loner	Einzelgänger
lookout for, to be on the	auf der Suche nach … sein
lost for words	nicht die richtigen Worte finden
loyalty	Loyalität
M	
managed	geleitet, verwaltet
management trainee	Trainee
mandatory	zwingend vorgeschrieben
manners; good ~	gute Manieren
marital status	Familienstand
market leader	Marktführer
master something	etwas meistern
master the business process	den Geschäftsprozess beherrschen
mature	reif
mediate	schlichten; vermitteln
mediation	Schlichtung; Vermittlung
mediator	Schlichter:in; Vermittler:in
medical certificate	medizinisches Attest
medical check-up	Kontrolluntersuchung
members of staff	Belegschaft
memory	Erinnerung
mention	erwähnen
mind your own business!	kümmern Sie sich um Ihren eigenen Kram
mind; have in ~	im Sinn haben
minimise	minimieren
miscellaneous	sonstiges

English	German
misconduct	Fehlverhalten
mistrust	Misstrauen
modest	bescheiden
monitor	überwachen
mood	Stimmung; Laune
motivators	motivierendes Element
multi-disciplinary	multi-disziplinär
mutual	gemeinsam, beiderseitig
myth	Gerücht; Mythos
N	
negotiate	verhandeln; aushandeln
negotiation	Verhandlung
negotiations; pay ~	Gehaltsverhandlungen
negotiator	Verhandlungsführer
nerve-wracking	nervenaufreibend
non-competition	Wettbewerbsverbot
normal hours	regelmäßige Arbeitszeit
notice	Kündigung
notice arrangement	Kündigungsregelung
notice in writing	schriftliche Kündigung
notice of dismissal	Kündigung
notice period	Kündungsfrist
notice; without ~	fristlos
nowadays	heutzutage
nutshell; in a ~	zusammenfassen; in aller Kürze; kurzum
nutshell; to put it in a ~	kurz gesagt
O	
O level/GCSE (= General Certificate of Secondary Education)	Mittlere Reife
obligation	Verpflichtung

English	German
obligations; legal ~	gesetzliche Verpflichtung
obliged	verpflichtet
obliged; to be ~	verpflichtet sein
observation	Beobachtung
observer	Beobachter
obtained	erhalten
obvious	offensichtlich
obviously	offensichtlich
occasion	Anlass
occasion; on the ~ of	anlässlich
offence	Verstoß
offend	verletzen
offender	Angreifer; Täter
off-hand	ungezogen; unverschämt
on behalf of	im Namen von
one-page report	1-seitiger Bericht
one-page summary	1-seitige Zusammenfassung
one-sided	einseitig
on-site	vor Ort
open dialogue	offener Dialog
open shop	Betriebe, in denen es den Beschäftigten freigestellt ist, ob sie Mitglied einer Gewerkschaft sein möchten.
openness	Offenheit
opinion	Meinung
opt for	sich für etwas entscheiden
optimise	optimieren
originated	hervorgebracht
out of date	veraltet
outgoing	aufgeschlossen
outline	darlegen, erläutern

English	German
outlook	Ausblick
outstanding	außerordentlich, herausragend
overall	insgesamt
overcome	bewältigen; meistern
overload	überladen
overseas qualifications	ausländische Qualifikationen
overtime	Überstunden
P	
panel interview	Interview mit mehr als einem Fragesteller
participate	teilnehmen
part-time	Teilzeit
peak; reach a ~	einen Höchststand erreichen
peer	Gleichrangige(r)
penalty	Konventionalstrafe, Strafe
pension scheme	Altersvorsorge
perception	Auffassungsgabe
performance	Leistung
performance agreement	Leistungsvereinbarung, Leistungsbewertung
performance targets	Leistungsziele
perks	Sonderzulagen
permit	erlauben
persistent	anhaltend
personnel development	Personalentwicklung
pick on somebody	auf jmdn. herumhacken
pick sth.	etw. auswählen
picketing	Streikposten aufstellen
pioneer	Vorreiter sein
pioneered	Pionierarbeit geleistet, Vorreiterrolle gespielt
place of jurisdiction	Gerichtsstand
place of performance	Erfüllungsort

English	German
placement	Praktikum
pleasure	Vergnügen
point a finger	beschuldigen
poke one's nose into something	sich in etwas einmischen
pool of employees	Mitarbeiterpool
pose a danger	eine Gefahr darstellen
position	Stelle
posting	Versetzung, Entsendung
posture	Körperhaltung
praise	loben
precede; to something ~	etwas vorausgehen
predict	vorhersagen
prejudice	Vorurteil
prejudiced	befangen sein; Vorurteile haben
prescribe	vorschreiben
previous	vorherig
prioritise	Prioritäten setzen
privilege	Privileg
probationary period	Probezeit
proceedings	Vorgehensweisen
procrastinate	aufschieben
produced	produziert, erzeugt
productivity	Produktivität
proficient	kompetent, etw. beherrschen
profit-sharing scheme	Gewinnbeteiligungsmodell
progress	Fortschritt
project leader	Projektleiter
projected image	dargestellte o. gezeigte Bilder
promote	befördern
promoted	gefördert, vorangebracht

English	German
promotion	Beförderung
proposal	Vorschlag
pros and cons	Für und Wider
prospective	zukünftig, potenziell
prospective employee	potenzieller:e Mitarbeiter:in
prospects	Aussichten
prosperity	Erfolg, Wohlstand
protagonist	Protagonist; Akteur
punctual	pünktlich
punctuality	Pünktlichkeit
punish	bestrafen
punishment	Bestrafung
purchasing clerk	Einkaufssachbearbeiter
purchasing department	Einkaufsabteilung
purchasing procedures	Einkaufsabläufe
purpose	Zweck
purpose; business ~	Geschäftszweck
pursue a career in	einer Laufbahn nachgehen
push someone	jemanden unter Druck setzen
put it briefly	kurz gesagt
put it in a nutshell	kurz gesagt
Q	
quandary	Zwickmühle
quarrel	Streit; Disput
questionnaire	Fragebogen
R	
raise awareness	Bewusstsein schärfen
ran	betrieben, geleitet, geführt
rarely	selten
rate of pay	Lohnsatz

English	German
rating	Bewertung
recapitulate	rekapitulieren, wiederholen
recharge one's batteries	seine Batterien wieder auffüllen
recipient	Empfänger
recommend	empfehlen
recommendation	Empfehlung
record	festhalten, protokollieren
records; to take ~	Buch führen; Aufzeichnungen machen
recruit someone	jemanden anwerben
recruitment	Personalbeschaffung
recruitment agency	Personalagentur
recruitment channel	Personalbeschaffungsweg
reduced	reduziert
redundancy	betriebsbedingte Kündigung
redundant; make someone ~	aus betriebsbedingten Gründen entlassen
references	Referenzen
reflect	widerspiegeln
regulation	Anordnung, Bestimmung
reinforce	bestärken
relationship	Verhältnis
reliability	Verlässlichkeit
reliable	verlässlich
relocation	Umzug
relocation expenses	Umzugskosten
reminder	Erinnerung; Mahnung
remuneration	Vergütung
renewal	Erneuerung
reorganized	reorganisiert
rephrase something	etwas umformulieren
reputation	Ruf

English	German
requirement	Anforderung
resilient	belastbar
resolved	geklärt, gelöst, beseitigt
respective	jeweilig
respond	antworten
restrict	beschränken
restriction	Einschränkung
restructure	umorganisieren
result-oriented	ergebnisorientiert
resume	Lebenslauf (USA)
retire	in Rente gehen
retirement fund	Pensionskasse
review (v)	überprüfen; überdenken
review	Rückblick
revise	überarbeiten
reward	Anerkennung; Belohnung
reward, alternative	alternative Entlohnung
right to	Anspruch auf; Recht auf
rock bottom; hit ~	Talsohle erreichen
rostrum	Rednerpult
rough times	harte Zeiten
rumour	Gerücht
S	
sackable	entlassungswürdig
sacrifice	opfern
salary arrangement	Gehaltsvereinbarung
salary bracket	Gehaltsstufe
salary expectation	Gehaltserwartung
salary increase	Gehaltserhöhung
salary package	Gehalt inkl. Sonderleistungen

English	German
salary structure	Gehaltsstruktur
salary; basic ~	Grundgehalt
sales clerk	Verkaufssachbearbeiter
sales director	Verkaufsleiter
sarcasm	Sarkasmus
satisfaction	Zufriedenheit
schedule	planen, Termin ansetzen
school fees	Schulgebühren
scope	Umfang
screen	Bildschirm; Leinwand
scruffy	ungepflegt
search	Suche
secondary employment	Nebenbeschäftigung
secrecy; maintain ~	Stillschweigen bewahren
section leader	Gruppenleiter
secured	sichergestellt, gesichert
see eye to eye	einer Meinung sein
selected	selektiert, ausgewählt
selection	Auswahl
self-assessment	Selbsteinschätzung
self-confident	selbstbewusst
self-explanatory	selbsterklärend
self-starter	Mitarbeiter:in mit Eigeninitiative
sensible way	vernünftige Weise
set of rules	Regelwerk
set out	festlegen
set someone's mind at rest	jemanden beruhigen
settle in	sich einleben
severability clause	Salvatorische Klausel
severance package	Abfindungspaket

English	German
severance pay	Abfindungszahlung
share participation scheme	Aktienbeteiligungsplan
shared values	gemeinsame Werte
sharpen up	verbessern
shop steward	Arbeitnehmervertreter
short-list someone	jemanden in die engere Auswahl nehmen
sickness	Krankheit
side-step	Schritt zur Seite
skilful	geschickt
skill	Fertigkeit, Fachkenntnis, Geschick, Fähigkeit
slide	Folie
slogan	Slogan
social security	Sozialversicherung
socialise	Beziehung pflegen
solicit	bewerben
solved	gelöst
sound	sich anhören
speaker	Redner
specialist knowledge	Fachkenntnisse; Fachwissen
speech	Rede
sponsorship	Förderung
spread	verteilen
spreadsheet	Tabellenkalkulation
staff handbook	Mitarbeiterhandbuch
staff members	Belegschaft
staff recognition	Anerkennung der Belegschaft
stake	Anteil
start-up company	neu gegründete Unternehmen
statement	Aussage
statutory	gesetzlich

English	German
statutory entitlement	gesetzlicher Anspruch
statutory provisions	gesetzliche Bestimmungen
step down	Schritt zurück
stepping stone	Sprungbrett
strategic thinker	Strategischer:e Denker:in
streamline	rationalisieren
streamlined	rationalisiert
stress	betonen
stress-creating	Stress verursachend
stressful	stressig
strike action	Arbeitskampfmaßnahme
strike; go on ~	streiken
struggle	sich schwer tun
subject to	unterliegen
submissive	autoritätsgläubig; unterwürfig
submit	vorlegen; einreichen
subordinate	untergeordnet; nachrangig; Befehlsempfänger
subsidiary	Tochtergesellschaft
subsidised	subventioniert
successor	Nachfolger
successor policy	Nachfolgeregelung
suggestion	Vorschlag
suitable candidate	geeigneter:e Bewerber:in
sum up	zusammenfassen
summarise	zusammenfassen
summon	aufrufen, bestellen
superior (n)	Vorgesetzter:e
supervise	beaufsichtigen
supervise	beaufsichtigen; überwachen
supervision	Überwachung

English	German
supervisor	Vorgesetzter:e
supplier	Lieferant
supply chain	Lieferkette
surround	umgeben
T	
take advantage of something	einen Vorteil aus etwas ziehen
take sides	Partei ergreifen
take-home pay	Nettoverdienst
target deployment	zielorientierter Einsatz
target; achieve one's ~	sein Ziel erreichen
task	Aufgabe
tax advantage	steuerlicher Vorteil
technical clerk	technischer:e Angestellter:e
termination interview	Kündigungsgespräch
terms and conditions	Geschäftsbedingungen
terms of reference	Aufgabenstellung
testimonial	Arbeitszeugnis
testing agenda	Testprogramm
thorough	gründlich
thoroughness	Gründlichkeit
threaten	drohen
threaten someone	jemanden drohen
throughout	durchweg
tidiness	Ordnung
time-consuming	zeitraubend
timeline	Zeitrahmen
timely manner	fristgerecht
timetable	Terminplan
timing	zeitliche Planung
tool	Instrument, Werkzeug

English	German
top-down	von oben nach unten
topic	Thema
track; get off ~	vom Weg abkommen, aus dem Ruder geraten
track; keep of ~	Überblick behalten
track; stay on ~	auf dem richtigen Kurs bleiben
trade union	Gewerkschaft
trainee	Kaufmännischer:e Auszubildender:e
training	Ausbildung
transparency	Folie
treat employees fairly	Mitarbeiter:innen gerecht behandeln
treat unfairly	unfair behandeln
trial period	Probezeit
tricky	schwierig; heikel
trigger; to ~ something	etwas auslösen
trim	kürzen
trimmed	gekürzt
trivia	Belanglosigkeiten
trusted	vertraut
trustworthiness	Vertrauenswürdigkeit
turnover	Umsatz
U	
uncompromising	kompromisslos
union member	Gewerkschaftsmitglied
union representative	Gewerkschaftsvertreter
union; join a ~	einer Gewerkschaft beitreten
unsatisfactory	nicht zufriedenstellend
unstructured	unstrukturiert, chaotisch
update	aktualisieren
upset; get ~	sich aufregen
utmost respect	höchste Anerkenntnis

English	German
V	
vacancies	offene Stellen
vacant position	offene Stelle
vacation	Urlaub
vacation; take ~	Urlaub nehmen
valid	gültig
valuable	wertvoll
value	schätzen
varied	verschiedene
venture	Unternehmung
verification	Überprüfung; Bestätigung
verify	bestätigen
violate	verletzen
violation	Verstoß
vocational course	Ausbildungslehrgang
volatile	brisant; unberechenbar
volatility	Unbeständigkeit
voluntary bonus	freiwillige Zulage
volunteer	etwas ehrenamtlich tun, freiwillig tun
W	
wage	Lohn
wage rate	Stundenlohn
warm welcome	herzlich willkommen
weakness	Schwäche
well-being	Wohlbefinden
What a cheek!	Was für eine Frechheit!
whistle-blowing policy	Richtlinie zur Meldung von Verstößen
white-collar worker	Angestellter
widely used	weit verbreitet
wishy-washy	vage

English	German
witness	bestätigen; bezeugen
word of mouth advertising	Mund-zu-Mund Propaganda
work environment	Arbeitsumfeld
work permit	Arbeitserlaubnis
workforce	Belegschaft
working environment	Arbeitsumfeld
working hours	Arbeitszeit
Y	
years of service	Dienstjahre

Lösungen zu den Übungen

Modul 1: Anforderungsprofil und Stellenbeschreibung

Task 1

1. decision
2. unskilful
3. regular
4. non-specialist
5. indirect
6. unemployment

Task 2

1. incumbent
2. opportunity
3. qualifications
4. gap year
5. questionnaire

Task 3

1. promote
2. fill
3. put
4. top
5. level

Modul 2: Erfolgreiche Stellenanzeigen formulieren

Task 1

1. considered
2. see
3. word-of-mouth
4. degree
5. potential

Task 2

1. disability	2. inexperienced	3. disadvantage
4. unavailable	5. exclude	6. independent
7. sales/selling		

Task 3

1. service provider	2. opinion	3. headquarters/head office
4. Previous experience	5. knowledge of English	6. Deadline

Modul 3: Bewerbungen analysieren

Task 1

1. attendees	2. unemployed	3. achieved
4. successfully	5. necessarily	

Task 2

1. unemployment	2. irresponsible	3. lack of experience/inexperience
4. illiterate	5. unavailable	6. include
7. useless		

Task 3

1. covering letter	2. consultant/adviser/analyst	3. temporary
4. applicant/candidate/jobseeker	5. referee	

Modul 4: Arbeitszeugnisse und Referenzen analysieren und schreiben

Task 1

1. testimonial or reference	2. salutation	3. performance
4. know-how	5. to whom it may concern	

Task 2

1. reliability	2. education/further education	3. sad/unhappy
4. refer	5. redundant	

Task 3

1. impersonal	2. exclude	3. dissatisfaction
4. inapplicable	5. past	6. anti-social
7. unattractive		

Modul 5: Bewerber:innen interviewen und bewerten

Task 1

1. prospects	2. panel	3. prospective
4. record	5. degree	

Task 2

1. CV/resume	2. commercial clerk	3. summarise
4. ice-breaker	5. environment	

Task 3

1. about/on	2. for … of	3. in
4. into		

Modul 6: Assessment Center (AC)

Task 1

1. present	2. vary	3. interest
4. consider	5. reassure	6. specify
7. assess	8. appreciate	9. summarise
10. introduce		

Task 2

1. interesting
2. selected
3. general
4. individual
5. personal
6. variable
7. special/specialised

Task 3

1. consultant
2. appreciate
3. parts
4. compares
5. correctly
6. decision

Task 4

1. aid
2. audience
3. interview
4. assessor
5. improvement

Modul 7: Einführung neuer Mitarbeiter:innen

Task 1

1. induction
2. headquarters
3. first impressions
4. health & safety
5. grievance

Task 2

1. on
2. to
3. after
4. on
5. on

Task 3

Noun	Verb	Adjective	Adverb	German
provision	provide	provided	XXX	bereitstellen; versorgen
requirement	require	required	XXX	benötigen
permission	permit	permissible/permitted	XXX	erlauben
dismissal	dismiss	dismissed	XXX	entlassen
equality	be equal/equalise	equal	equally	gleich sein: Gleichheit
impartiality	be impartial	impartial	impartially	unparteiisch sein

Modul 8: Arbeitsvertrag erstellen

Task 1

1. oral/verbal	2. trial period/probationary period	3. tasks/duties
4. Payment	5. overtime	

Task 2

1. vacation	2. grant	3. superior
4. sick note	5. remuneration	6. termination/terminate
7. continuing	8. confidentiality	

Task 3

1. scope	2. remuneration	3. charge
4. collective	5. check-up	6. miscellaneous

Modul 9: Mitarbeiterbeurteilung

Task 1

1. colleague	2. target	3. performance
4. appraisee	5. agenda	

Task 2

1. in	2. for	3. on
4. from	5. in	6. about
7. on		

Task 3

1. successfully	2. different	3. typical
4. reviewing	5. taken	6. keep

Modul 10: Vergütung

Task 1

1. do more	2. private medical insurance	3. reflected
4. promoted	5. remuneration	6. subsidised

Task 2

1. perform	2. expectations	3. doctorate
4. personnel	5. department	

Task 3

1. for	2. with	3. by
4. to	5. by	6. At
7. for	8. for	

Modul 11: Tarifverhandlungen und Gewerkschaften

Task 1

1. restructured	2. security	3. achieve
4. significance		

Task 2

1. non-unionist	2. join the union	3. employ
4. unfairly	5. improved conditions	

Task 3

1. Let's talk about the matter point by point.
2. Very interesting, but we wish to include the following point.
3. Let's down tools!
4. Let's deal with formalities first.

Modul 12: Mitarbeitermotivation

Task 1

1. track	2. benefit	3. weaknesses
4. staff	5. approach	

Task 2

1. b	2. b	3. a
4. b	5. b	

Task 3

1. misunderstood	2. connection	3. unimportant
4. motivate	5. flexibility	

Task 4

1. shop	2. over the leadership	3. in … framework of the team
4. substitute for	5. share in	6. general
7. at … level	8. enlarge	9. stimulate
10. reorganise	11. depend on	12. increasing

Modul 13: Konflikte am Arbeitsplatz erkennen und lösen

Task 1

1. product, production, produce	2. harassment	3. clarity
4. disturbance	5. interruption	6. prevention
7. encouragement		

Task 2

1. documented	2. knowledge	3. impartial
4. sensitive	5. react	

Task 3

1. work conflict	2. apologise	3. complaint
4. distributed	5. colleague	

Modul 14: Entsendung ins Ausland

Task 1

1. a placement or a secondment
2. salary
3. overview
4. grievances
5. high potential

Task 2

1. reason
2. nevertheless
3. Deductions
4. my chest
5. setting up

Task 3

1. of
2. for
3. out of
4. on
5. to

Modul 15: Präsentationen und Berichte

Task 1

1. check
2. expert
3. opportunity
4. concludes
5. slide
6. aid

Task 2

verb	noun	adjective	German
guide	guide	guiding	führen
visualise	visualisation	visualised	visualisiert
compare	comparison	compared	vergleichen
introduce	introduction	introductory	einführen
find	findings	found	finden; Erkenntnisse
be punctual	punctuality	punctual	pünktlich
donate	donation	donated	spenden
continue	continuation	continuous	fortführen

Task 3

1. I have been working/I have worked at ICS as a consultant for three years.
2. Today I'd like to look back at the last fiscal/business/financial year.
3. Let me say from the outset/beginning that ...
4. I'm happy to answer any questions.
5. May I draw your attention to the handouts in front of you.

Modul 16: Reden zu Firmenanlässen

Task 1

1. raise/future	2. rise/toast	3. propose/toast/conclusion
4. suggest	5. celebrate/agreement	

Task 2

1. frankness/honour	2. joyful/task	3. pretend/rare
4. join/prosperity		

Task 3

1. on	2. over	3. at
4. in	5. to	6. on
7. in	8. on	

Modul 17: Die Abmahnung

Task 1

1. workplace	2. in writing	3. point a finger at/accuse
4. behaviour	5. head of department	

Task 2

1. clearly	2. differently	3. dissatisfied
4. improvement	5. procrastinate	

Task 3

Noun	Verb	Adjective	Adverb	Opposite adjective
fairness	be fair	fair	fairly	unfair
improvement	improve	improved	XXX	not improved
specification	specify	specified	XXX	non-specified
difference	differ	different	differently	indifferent
surprise	surprise	surprised	surprisingly	not surprising
possibility	make possible	possible	possibly	impossible

Modul 18: Die Kündigung

Task 1

Noun	Verb	Adjective	German
1. termination	terminate	terminated	kündigen
2. renewal	renew	renewed	erneuern
3. advice	advise	advisable	raten
4. involvement	involve	involved	involvieren
5. expectation	expect	expected	erwarten
6. dismissal	dismiss	dismissed	entlassen
7. procedure	proceed	XXX	fortfahren, verfahren
8. receipt	receive	received	erhalten

Task 2

1. followed
2. package
3. serious
4. remind
5. entitled

Task 3

1. to
2. of
3. from
4. in
5. about
6. during
7. against
8. for

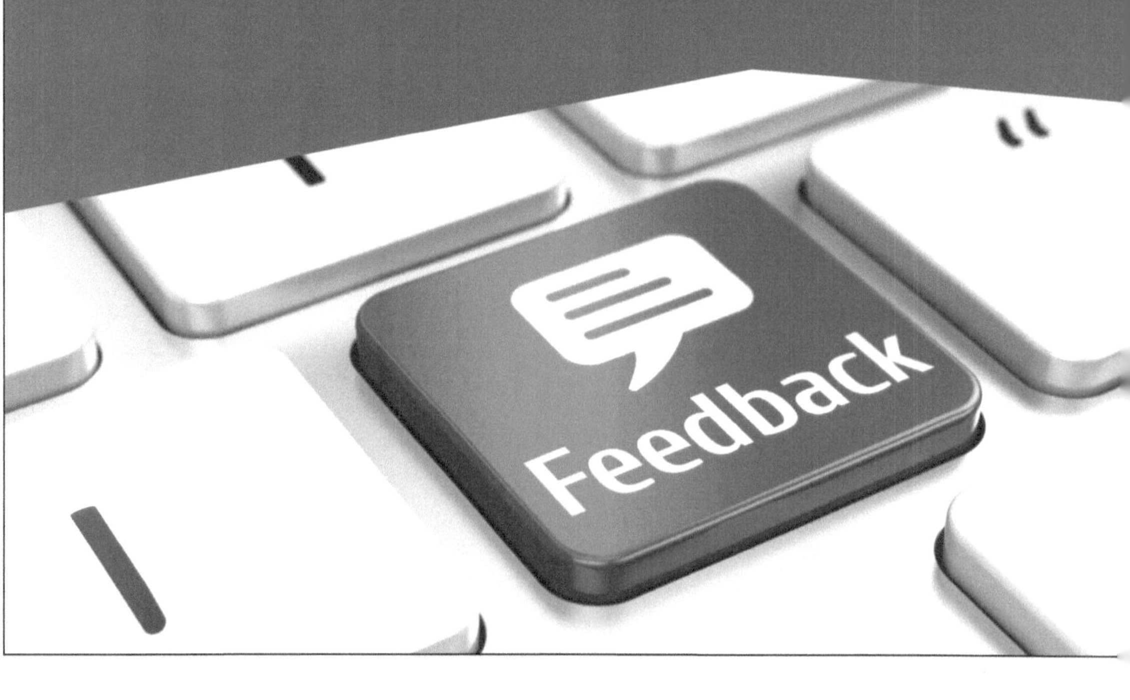